EXCAVATION & GRADING HANDBOOK

by
Nick Capachi

Craftsman Book Company
6058 Corte del Cedro / P.O. Box 6500 / Carlsbad, CA 92018

The author wishes to thank **Caterpillar Inc.** of
Peoria, Illinois, for providing photographs of Caterpillar
Paving Products for use in this book.

Looking for other construction reference manuals?

Craftsman has the books to fill your needs. **Call toll-free 1-800-829-8123**
or write to Craftsman Book Company, P.O. Box 6500, Carlsbad, CA 92018 for
a **FREE CATALOG** of over 100 books, including how-to manuals,
annual cost books, and estimating software.

Library of Congress Cataloging-in-Publication Data

Capachi, Nick, 1934-
Excavation and grading handbook.

   Includes index.
   1. Excavation.   2. Road construction.   3. Earthwork.
I. Title.
TA730.C28   1987   624.1'52   87-30570
ISBN 0-934041-29-6

First edition ©1978 Craftsman Book Company
Second edition ©1987 Craftsman Book Company

Ninth printing 2001

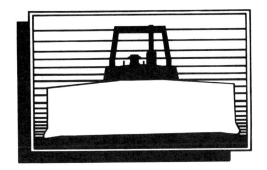

Contents

Understanding Road Survey Stakes

This manual is a practical guide to excavation and grading. It's written for anyone who has to plan, estimate or supervise excavation and compaction for building sites, highways, drainage channels or trenching. It also covers installation of water, sewer and drain pipe and the laying of asphalt concrete pavement.

The first edition of this manual was published in 1978. Since then the book has gone through five printings and has been adopted as the primary reference by many schools and in many apprentice training programs. This second edition reflects the changes we've seen in the excavation business since the mid-1970's. Equipment has changed, materials have changed and the best way of getting the job done has changed. I had to make changes in every chapter of the first edition to be sure this second edition describes good current practice for excavation and grading contractors. In many cases I'm indebted to readers of the first edition who made suggestions for improvements or recommended alternate procedures. This second edition also includes new chapters on using contour line drawings to control excavation, using laser levels, and trench compaction.

I'm going to start by assuming that you're new to excavation and grading. The first three chapters cover the basics: reading and following survey stakes, understanding excavation plans and how excavation contractors use contour line drawings. If you've been working in the excavation and grading business for a while, you'll probably be able to skip the first few chapters. But if you need information on plan reading and stake markings, it's here for your use.

Now, let's get started at the beginning, reading and following survey stakes.

Survey Stakes

Excavation for roads, buildings and pipelines begins with a survey of the area where the excavation will be done. A survey crew working for the engineering firm that's designing the project will set out stakes and *hubs* which identify points that are on the construction plans. When a precise distance or elevation is needed, a surveyor's tack on top of the hub establishes the point from which elevations and distances are measured.

Beside each hub there will be an *information stake* which explains in surveyor's code the grades at various distances from the hub or other reference stake or point. It's essential that you know how to read the markings on these information stakes and follow the instructions they provide.

Figure 1-1 shows the kind of markings you'll find on an information stake. Usually these are called *cut* or *fill* stakes, depending on the type of excavation required. The front, back and both sides of a cut stake are shown in Figure 1-1. Below the stake there's a cross section drawing of the existing grade and final road grades that are described on the stake. This drawing will help you understand the markings on the stake. Refer to the drawing as I explain the markings on the information stake in Figure 1-1.

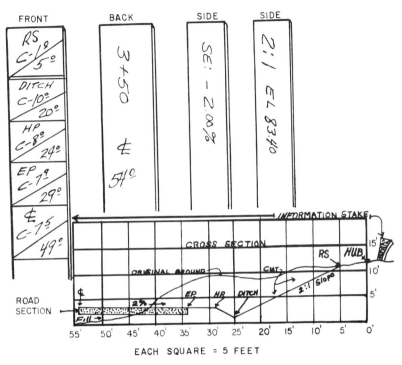

Cut stake reading
Figure 1-1

Cut Stakes

Look first at the stake labeled *front* in the upper left of Figure 1-1. That's the front of the information stake. The *RS* at the top of the stake means that there's a reference stake to be established, and that reference stake is the point from which measurements and elevations are taken. Find the reference stake in the drawing. It's labeled *RS* and is just to the left of the hub at the right edge.

Below the letters *RS* on the stake, you see the letter *C* and a dash followed by some numbers. Below that you see a diagonal line and some more numbers. These markings above and below the diagonal line identify the amount of cut and distance needed to establish the correct grade at the reference stake. The number above the line is the elevation and the number below the line is the distance. In this case, the reference stake shows a *cut* 1 foot and zero tenths of a foot below the level of the surveyor's hub, 5 feet and zero tenths of a foot from the hub.

Some surveyors may use *RP* instead of *RS*. *RP* means reference point. Treat it exactly the same as the *RS*.

Notice that distances and elevations are measured in feet and tenths (or hundredths) of a foot, not feet and inches. The small number above the small horizontal line shows decimals of a foot. That's a little different from what you're probably used to, but you'll appreciate the difference when adding and subtracting feet and decimals of a foot rather than feet, inches and fractions of an inch. I'll explain more about this measuring system, called engineer's measure, later in this chapter.

The two horizontal lines below the first set of measurements are very important. All measurements above the double horizontal line are taken from the hub beside the information stake. The double horizontal line means *and then,* indicating that all measurements and elevations from that point down on the stake are taken from the RS point and not the hub. Note this very carefully: if the double horizontal line were replaced with a single horizontal line, all measurements and elevations would be taken from the surveyor's hub rather than a reference stake established.

The next information on this stake shows the elevation and location of the ditch cut. It's to be 10 feet lower than the RS point and 20 feet from it. The grade falls 10 feet over a horizontal distance of 20 feet, thus creating a 2:1 slope. You'll see this indicated on the drawing. For every foot of cut, the grade line moves horizontally 2 feet.

Notice that all measurements are made from the reference stake. The ditch is cut 10 feet below the reference stake and 20 feet from that stake. Also note that the 20 foot distance is measured horizontally, not diagonally, from the reference stake. Glance at the drawing to be sure you understand how the 20 foot distance to the ditch is measured.

The next reading is the *hinge point* (HP) grade and distance. Note the hinge point on Figure 1-1. It's cut 2 feet above the ditch cut. The HP information indicates the grade must come up 2 feet and move out 4 feet. By computing the amount the HP rises from the ditch and the distance it moves towards the center of the road, you can see that it's again a 2:1 slope. Reading down the information stake, the next grade and distance is the *edge of pavement* (EP) point. The grade will be 7.9 feet below the reference stake. Notice the cut at EP is 0.10-foot less than the HP. The reason for this is that the road grade rises 2% in the 5 feet from HP to EP. Multiplying 5 feet by 2% gives the amount the shoulder rises in that distance (5.00 x .02 = .10).

The next markings give the centerline cut. You can see that the cut is again less than the previous cut at EP. Subtracting the 29 feet at EP from the 49 feet to the centerline leaves 20 feet. So the centerline is 49 feet from RS and 20 feet from EP. The cut at the centerline is 0.40 foot higher than EP, giving a 2% slope from the centerline to EP. This is computed by multiplying the 20 feet by 2% (20.00 x .02 = .40).

Look at the back of the cut stake. It's marked 3 + 50, indicating that this station is 350 feet from station 0 + 00, the point from which the survey began. Below the station number is the distance from the surveyor's hub to the center of the road. This includes 5 feet to the RS and 49 feet from the RS to the centerline, a total of 54 feet.

Now note the first stake marked *side*. This side of the stake identifies the percentage of slope from the centerline to HP. The minus sign indicates that the centerline slopes down to the HP. If it were a plus sign instead, the centerline would be sloping *up* to the HP. The second stake marked *side* first gives the rate the cut slope falls from RS to the ditch. In this case, it is 2 feet out for every foot downward. The second group of numbers is the elevation of the hub above sea level.

Comparison of inches and decimals of a foot— Notice that all measurements are in feet and tenths of a foot rather than feet and

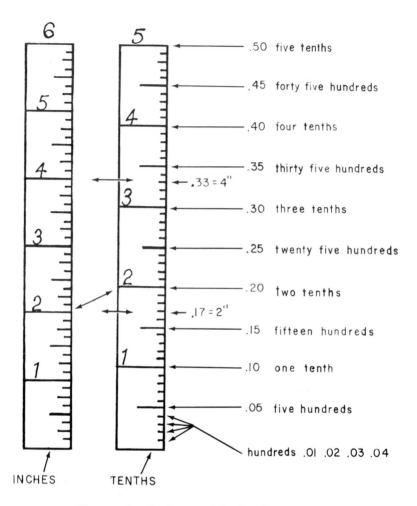

Comparing inches and decimals of a foot
Figure 1-2

inches. Setting grades requires many additions and subtractions. Using decimals speeds the work and makes errors less likely. Figure 1-2 compares inches with decimals of a foot.

If you're uncomfortable reading distances in tenths and hundredths of a foot, think of one foot as being like a dollar bill. One dollar is the same value as 100 pennies; one foot is the same distance as 100 hundredths of a foot. One dollar is the same value as 10 dimes; a foot is the same distance as 10 tenths of a foot. Pennies are hundredths. Dimes are tenths or ten hundredths.

Fill Stakes

Figure 1-1 shows a cut stake where material must be excavated to reduce the existing grade to the finish grade. Figure 1-3 shows a typical fill situation where soil has to be deposited to build up the existing grade. Again, the illustration shows four sides of the stake and the road cross section. The *RS* means that the reference stake (to the right of the hub) is the starting point and the place from which all measurements and grades are measured. The cuts and fills given for the RS point will be measured from the hub. Here, the RS is located 1.8 feet above the hub and 3 feet from it. The grade setter will have to set the reference stake the indicated horizontal distance from the hub and draw a horizontal line on the stake at the elevation given on the surveyor's information stake. Or he may elect to place his own hub there.

Reading down the stake, the two horizontal lines mean *and then,* indicating that the grade setter must measure from the RS point for the next fill and distance given instead of measuring or shooting grades from the original hub. For the *hinge point* (HP), measure 10 feet from the RS hub or lath. At this point a fill of 5 feet must be made to obtain the required grade. The hinge point is the place where the fill slope stops and the road grade begins. It's sometimes called the *catch point.* Next, reading down the stake, is the EP. This is the edge of the pavement and has a F-5.12 fill 14.0 feet from RS.

Below the EP data is PG. This is the *projected centerline grade.* In most cases, the surveyors will mark it as the centerline and not PG. From the RS, measure 32 feet and fill 5.66 feet. This will put the PG or centerline 18 feet from the EP and 0.54 of a foot higher.

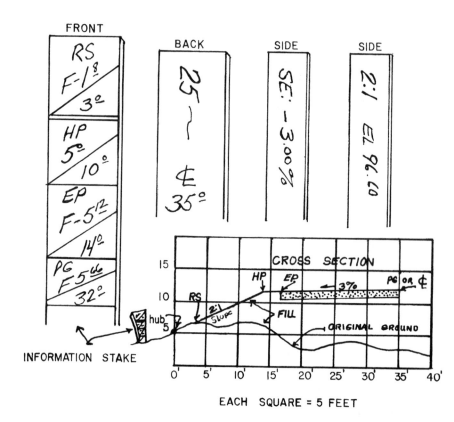

Fill stake reading
Figure 1-3

The stake marked *back* has a 25 and a little worm line standing for +00. Some jobs might have an A, B, C line with one being at the centerline. Check the plans to learn what these lines mean. The 25 + 00 identifies this stake as being 2,500 feet down the line from the point where the measurements started. The point the surveyors start from is most likely marked 0 + 00, but may not be in all cases.

Next, reading down, notice a C and an L, one over the other. This means *centerline.* The number 35 below that means that the center of the road is 35 feet from that point. Look back to the stake marked *front* and notice that when the RS distance of 3 feet is added to the PG distance of 32 feet, the total is 35 feet, the same distance as marked on the back.

The stake labeled *side* is marked SE-3.00%. This is the percentage that the road bed slopes from the centerline to the hinge point. On the far right stake marked *side,* the first reading is 2:1 (two to one). This is the rate the fill slope will rise from RS to HP. Notice that the first stake has a 5 foot fill over a 10 foot distance. This is what the 2:1 indicates. The next item down the stake is EL 96.6. This is the elevation of the hub at the information stake. All cuts or fills were computed from that hub by the surveyors.

What I've described so far in this chapter is more or less standard procedure for indicating elevations and distances on road stakes. However, surveyors in some counties and cities follow slightly different procedures. Some surveyors provide more information on the stakes. The stakes in Figure 1-4 show what you might see on some county or city road stakes.

The top of the stake has a 2 with a circle around it. This indicates that the first cut starts 2 feet out. The next markings indicate that the cut is 4 feet at 10 feet from the stake. The slope will again be 2:1 because the first 2 feet are not cut and the cut over the next 8 feet is 4 feet. Look at Figure 1-4 again. Notice that there is no double *and then* line. This means that you must take all measurements and grade shots from the hub set by the surveyors rather than from an RS or RP point as on the previous stakes shown.

Reading down the stake at the left in Figure 1-4, we find a second group of numbers that shows the top of the shoulder cut. This was referred to as the HP, or hinge point, on previous stakes. Notice there's no EP distance or elevation on the stake. In this case, you must look at the plans for the distance from shoulder to edge of pavement, and the elevation.

Engineering companies follow different conventions when marking their stakes. But the plans should clarify what is intended and which points are actually indicated. If something isn't clear, don't guess. Call the engineering company that created the drawing and marked the stakes. They should be eager to help.

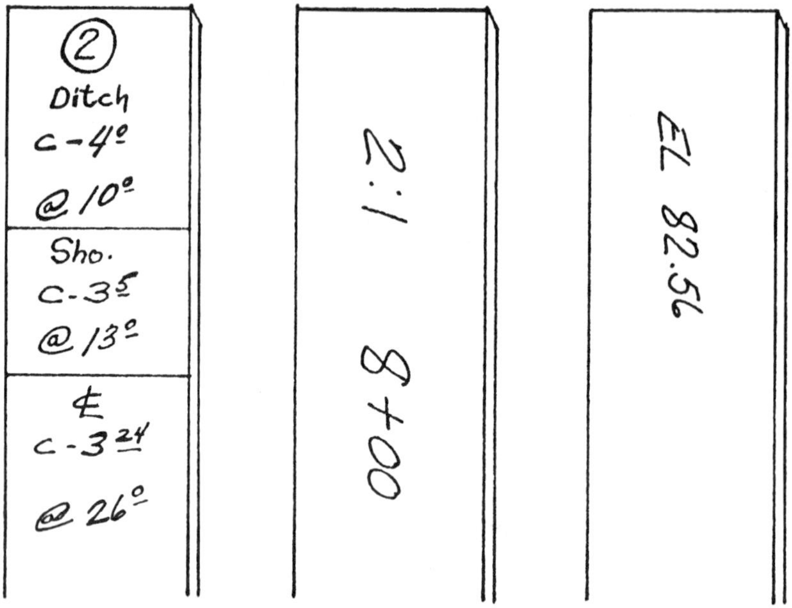

Some county and city stake readings
Figure 1-4

The second illustration in Figure 1-4 is the back of the stake. It shows the rate of fall of the cut slope (2:1) and the station number (8 + 00). The far right illustration is the side of the stake. It gives the elevation above sea level (EL 82.56). In some cases the hub elevation will not be on the stake at all. It may be replaced with the percentage of slope on the road or both may be omitted entirely. The back of the stake in Figure 1-4 will still have the station number but no centerline distance because all the front measurements are from the hub and not an RS or RP point.

Many stakes have very little information. They have just the details required to allow you to set the grades. These stakes always have the station number on the back, though the percentage of slope and hub elevation may be absent.

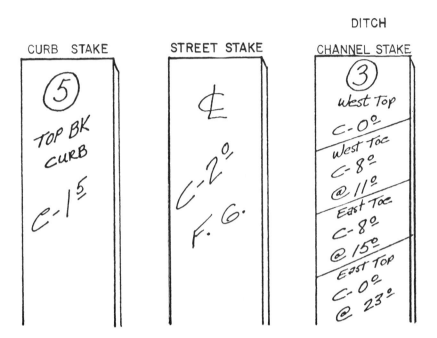

Miscellaneous information stakes
Figure 1-5

Miscellaneous Information Stakes

Curb stake— Note Figure 1-5. The stake at the left is what you'd expect the surveyor to set for cutting and setting curb grades. From the hub at the base of this information stake, move out 5 feet and down 1.50 feet to the top of the curb to set the curb forms.

In some cases, the surveyors will give the front lip grade or even the flow line grade. If not, you'll have to determine the distance from the back of the curb to the lip. This information is available in the plans. When setting curb subgrade, determine the thickness of the curb plus any aggregate base, if it's called for under the curb. The thickness of one or both must be added to the cuts and subtracted from the fills to find the subgrade rather than the finished grade level.

Street stake— The center stake in Figure 1-5 is a stake you would expect to find in a subdivision for the first road cut. The front of the stake indicates the centerline of the street, and the cut or fill to the finished grade. In this case, there's a 2 foot cut to the finished grade (F.G.). The plans should show the road width, percentage of slope or crown, and the thickness of the road section. Remember to add the thickness of the road to this cut. The station number will be on the back of the street stake.

Ditch channel stake— The stake at the far right in Figure 1-5 is a grade stake for a ditch or small channel. The 3 in a circle is the distance from the hub where the first cut starts. The *west toe grade* indicates the first slope and the bottom of that slope. The *east toe* is the bottom of the slope on the opposite side of the ditch. Both toe cuts are the same, so the bottom is flat. The *east top cut* is where the cut will be started on the opposite side. Subtracting the 3 foot offset from the 23 feet distance to the east top cut gives the distance across the top of the ditch, 20 feet. Subtract the small toe distance from the larger. This gives the width of the ditch bottom, 4 feet.

To find the rate of slope from the top cut to the toe of the channel, subtract the distance given to the top cut from the distance given to the toe cut. The 3-foot offset must be subtracted from the west side distance of 11 feet. This will make the distance 8 feet from top cut to toe on each side. Dividing the cut of 8 feet into the 8 foot horizontal distance gives an answer of 1. This indicates that for every foot cut vertically, the slope moves out 1 foot horizontally. That's a 1:1 slope.

A stake with only a few markings will usually provide all the information you need to do the excavation. If something is still unclear, the plans should have the answer you're looking for.

In this chapter we've described grades by either a ratio of run to rise, or as a percent above the horizontal. Most grades in excava-

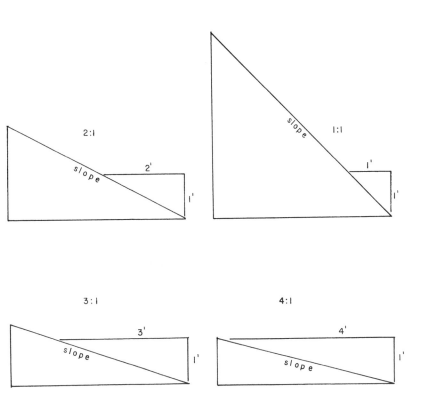

Slopes 1:1 to 4:1
Figure 1-6

tion work are expressed as a ratio of horizontal distance (run) to vertical distance (rise). Figure 1-6 illustrates the four most common slope ratios, and should help you visualize most of the slopes you work with in excavation.

If you're still confused about the work required after reading the surveyor's stakes and checking the plans, ask the survey crew about it if they're still on the job. If they've left, call the engineer and have him clarify the problem or send the survey crew out for a field meeting. *Be sure you know what's required before beginning the work.*

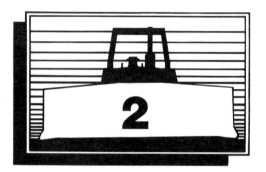

Plan Reading

Chapter 1 explained how to read survey stakes. The markings on survey stakes are a shorthand way of expressing what's on the excavation plans. As I explained in Chapter 1, you'll often have to refer to the plans to fully understand what work must be done.

This chapter will explain the lines and symbols you'll find on excavation plans for subdivisions and highway projects. It's essential that the foreman and grade setter be able to read and understand the plans. It's also an asset to your company to have some operators and laborers who can read plans. Any time the surveyors use an unfamiliar abbreviation or notation on a stake, the foreman or grade setter will have to check the plans to see what it means.

Subdivision Plans

Figure 2-1 is the first sheet of a set of plans. It shows a typical street cross section plan that you might find on a subdivision or road excavation job. Notice that every distance is given from the

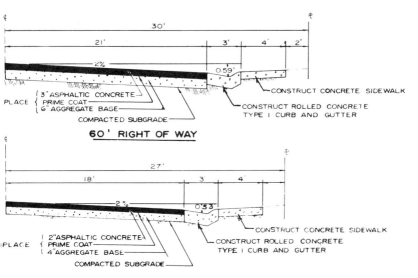

60' RIGHT OF WAY

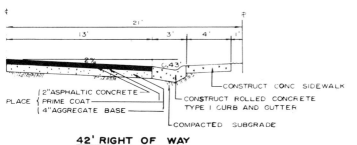

54' RIGHT OF WAY

42' RIGHT OF WAY

SHEET INDEX

Typical street cross section
Figure 2-1

centerline; to the front of the curb, to the back of curb, to the back of the walk, to the property line. The number at the top center of each street cross section is the distance from the centerline to the property line. The lower group of figures indicates the distance from the centerline to the lip of the curb, the width of the curb, the width of the sidewalk, and the distance from the back of the walk to the property line.

Refer to the top street cross section. Above the street surface and curb, the figures 2% and 0.59 are shown. The figure 2% is the slope of the street surface from the centerline to the lip of the curb. The figure 0.59 indicates that the flow line of curb and gutter is fifty-nine hundredths below the centerline of the street.

Below the cross section drawing, the type of paving material and the thickness required are indicated. The actual type of asphaltic concrete, prime coat, aggregate base, curb and gutter and sidewalk required will be described in the job specifications. In some cases, the job specifications will refer to a master set of specifications. The percentage of compaction required will also be spelled out in the specifications.

The right-of-way distances *(60', 54'* and *42')* are the measurements from property line to property line.

The cross section indicates a "type 1" curb is required. The dimensions of this curb and walk will be explained in some county, city, or state standard specifications or drawings.

Below the cross section drawings is a sheet index. This index shows the title of each drawing in the complete set of plans and gives the page number for each drawing. Every large set of plans has an index. Every set of plans will have a typical cross section drawing, whether it's a subdivision street with curbs or a highway with ditches at each edge. This plan is usually referred to as the "typical." The typical drawing usually shows only half of each street. That's all you need when both halves are exactly the same.

The Grading Plan

The grading plan in Figure 2-2 may at first glance seem to be very complicated. Actually it's fairly simple to understand. This is a portion of a grading plan for a subdivision. It supplies all the in-

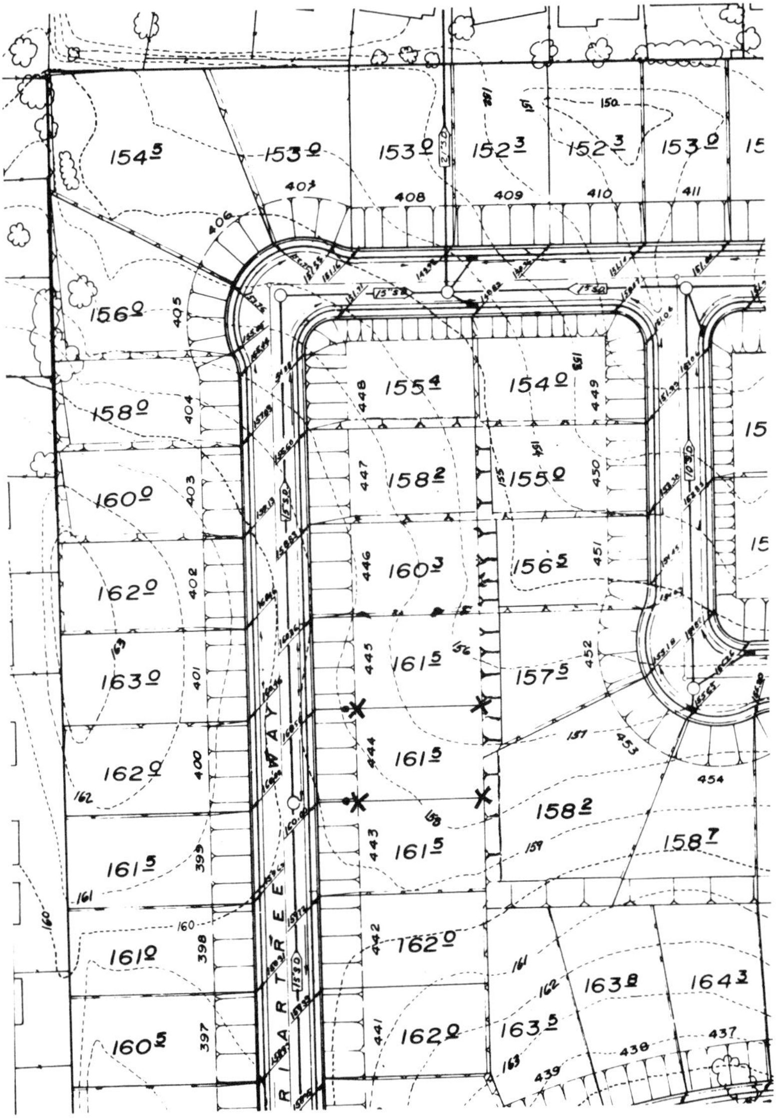

Grading plan
Figure 2-2

formation needed to grade the lots. Each lot is numbered and has a finished lot pad elevation indicated. The lot numbers are the smaller numbers from 397 to 454.

Notice lot number 444, slightly to the left of center in Figure 2-2. The elevation of the building pad is 161^5. This indicates that the pad is one hundred sixty-one feet and five tenths above sea level. The larger the elevation number, the higher the pad. Lots 443 to 445 are level. Lots 446 to 448 get progressively lower. Lot 448 is six feet one tenth (6^{10}) lower than lot 445. To find the difference in lot elevations, subtract the smallest elevation from the largest elevation.

Lot 444 has X's marked in each of four corners. Notice also two dots in front of the two front lot X's. The surveyors will set hubs and stakes where the two dots are marked. These hubs will be 5 feet from the X's which mark the front lot pad corners. Each stake will have a 5-foot offset marked on it, and indicate the amount of cut or fill required for the spot where the X is marked on the plan. The back lot stakes are usually set at the rear lot corners where the back X's are marked. Usually no offset will be marked on them.

The elongated Y's at the front and back of lot 444 indicate that the ground slopes down from the lot pad to the back of the walk in front, and to the lower lots at the back. Notice the same elongated Y's at the sides of each lot that is above or below the adjoining lot. This helps you visualize the slope.

There are several small elevation numbers in the street area. These elevations indicate the level at the top back of the sidewalk.

Contour lines— The dotted lines that wander through the grading plan indicate the elevation of the existing ground. They're called *contour lines*. Each contour line has a number somewhere along it. This is the elevation above sea level. Each line is at the given elevation no matter how many turns it makes or how far it goes.

The contour lines can tell you approximately how much cut or fill is needed on any lot or in the street, without checking the cut or fill indicated on each lot or street stake. Notice the contour lines that run through lot 443. The top contour line has an elevation of 158 feet. The bottom contour line, if followed to the left to lot 453, is at elevation 159. The finished lot pad elevation is indicated as 161^5. Subtract each of the two contour elevations from the finished lot pad elevation. You see that a fill of three feet five tenths (3^{50}) is required at the top contour line and a fill of two feet five tenths (2^{50}) is required at the bottom contour line.

Check the contour lines running through lot 453. You'll notice that fill is needed at the top of the lot pad and a cut must be made at the bottom of the lot pad. The center of the lot is near the finished lot pad elevation before excavation begins. You could color all the cut areas one color and the fill areas another color. This would help you see the fill and cut areas easily and would be useful in developing an excavation plan.

Computer programs are now available that will convert cut and fill data into a plan drawn on a 25-foot grid. The computer plotter creates a plan showing cuts and fills in different colors. Streets, building pads, and parking area perimeters are shown clearly. A plan like this makes it easy to anticipate problems and get maximum productivity from the equipment. The foreman could study the computer printout the night before starting work and have excavation procedures worked out before ever seeing a stake on the project.

Plan and Profile Sheets

There will be a plan and profile sheet similar to Figures 2-3 and 2-4 for every street in the project. The plan and profile sheet adds information not given on the grading plan or the street cross section drawings. Figure 2-3 is the profile of a 500-foot section of Zenith Drive, including a plan for the sewer and drainage lines. Profiles are drawn on graph paper. In this case, each square represents one foot vertically and 10 feet horizontally.

Going up the right side of the graph paper are figures from 140 to 170. These figures represent the elevation above sea level. At the bottom of the graph are figures from $29 + 28.78$ to $33 + 56.32$. These are the stations indicated along the centerline of the plan drawing, Figure 2-4.

At the top right corner of the graph paper, you'll see the centerline symbol and the words "finished grades." This means

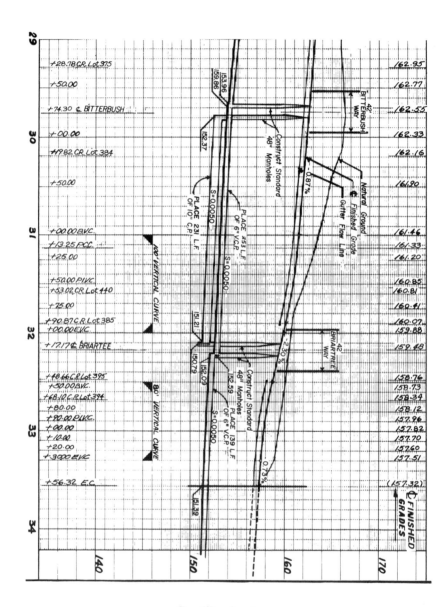

Profile sheet
Figure 2-3

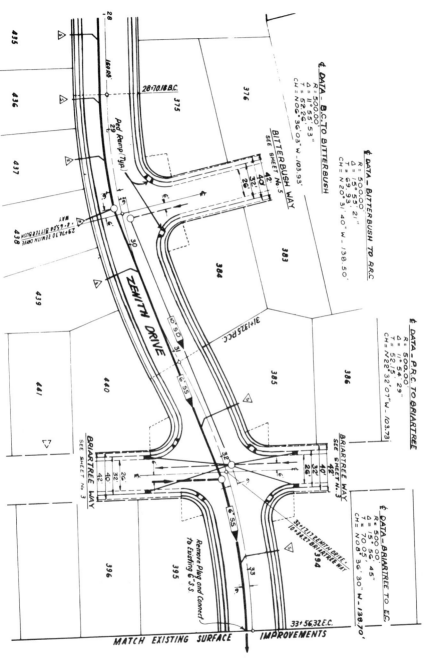

Plan sheet
Figure 2-4

that all the figures to the left of it, along the top of the page, are
finished road centerline elevations. At the bottom of the graph
paper, written horizontally, are the precise station numbers those
finished elevations relate to.

Find the words "natural ground," "finished grade" and "gutter flow line" in the upper left quarter of Figure 2-3. The first line
says "natural ground" and has a line and arrow pointing to an irregular horizontal line. This indicates the original ground elevation
before any excavation has been done. The second line has the
centerline symbol and the words "finished grade." The line and
arrow point to a heavy horizontal line. This indicates the elevation
of the finished centerline of the street surface.

Notice that the slope of the centerline is indicated in three spots
where the slope changes. These are -0.87%, -2.30% and -0.73%.
Along the centerline finished grade line, there are four small
circles. Follow down the vertical graph lines that split those circles
until they reach the black triangles. Between each set of black
triangles, a distance is given (100 feet and 80 feet) and the words
"vertical curve" appear. This indicates that between those
triangles and for the distances given, the road centerline curves.
You can see in Figure 2-3 that the curves are in different directions. One is convex and the other is concave. The portions of the
centerline finished grade line that are straight are labeled with the
appropriate percentage of slope. The straight areas are called
tangents.

The lighter line just below the centerline finished grade is the
gutter flow line. In Figure 2-3, this line follows the centerline
finished grade exactly, with the same tangents and vertical curves.
This is not always the case. Many times the curb flow line will not
parallel the centerline finished grade. When the gutter line does
not parallel the finished grade centerline, there will be another
group of figures just below the centerline finish grade which indicate gutter flow line elevations.

Locate the four heavy vertical lines on Figure 2-3 that have "42'
Bitterbush Way" and "42' Briartree Way" written between them.
These show that two side streets intersect Zenith Drive. All the
lines below the gutter flow line indicate pipe below the street sur-

face. The vertical lines which are close together and come to a point at the centerline finished grade are manholes. Note the instructions "Construct Standard 48" Manholes." There are one or two sets of figures at the bottom of the vertical lines that represent manholes. These indicate the flow line elevations of the pipe coming into the manhole and the elevation of the pipe leaving the manhole.

The horizontal lines at the bottom of the manhole indicate the pipe and direction of flow that intersect the manhole. The top two horizontal lines represent the sewer line, and the wider bottom two lines show the drain line. Notice the writing with arrows pointing to the sewer and drain lines. "Place 253 LF of 6" V.C.P." means that, between the two manholes, there will be 253 feet of 6" vitrified clay pipe. The writing directly below says "Place 231 LF of 10" C.P." This means that, between the two manholes, there will be 231 feet of 10" concrete pipe.

The plan gives the percentage of slope along each pipe section. In this case, the slope is indicated as "S = 0.0050. Given the rate of slope for the pipe, street, and flow line of the gutter, you can compute the amount of slope per foot. This is done by multiplying the percentage of slope by the total length. In this case: 0.0050 x 253' = 1.265'. The sewer line would drop 1.265 feet in 253 feet. Now divide that by the length (1.265' ÷ 253' = 0.005'). It drops 0.005 feet in each foot of length.

Staking the curves— Look now to Figure 2-4. This is the plan drawing of the profile in Figure 2-3. Find the small station numbers from 28 to 33 along the centerline. Notice also the dotted and solid lines running perpendicular through the centerline, with a small circle where the two lines meet. At the far left you'll see "28 + 70.18 B.C." to the left of the line. The B.C. is the abbreviation for *begin curve.* You'll see 33 + 56.32 E.C. at the far right side and at the top of the street. E.C. is the abbreviation for *end of curve.* A station number is written on a line between the B.C. and E.C.; 31 + 13.25 P.C.C. This indicates *point of curve change.*

There are five circles with station numbers at each. These are distances the surveyors must know to stake the curve. Notice the four groups of centerline data written across the top of the sheet. The data supplies the surveyors with the distances and transit readings needed to turn the correct angles to stake the centerline curve. No radius point is set for a curve of this type. There will be radius points set at each corner where the two dotted lines on each corner meet.

The plan drawing shows the lot location and numbers but doesn't give lot elevations as found in the grading plan, Figure 2-2.

Sewer mains and storm drain lines— The plan drawing of Zenith Drive shows a sewer main and a storm drain line in the street section. There are three pointed box symbols written in the street area that indicate the size of the pipe and the direction of flow. The symbol at the left marked *10" S.D.,* means that the pipe has a 10" inside diameter and is a storm drain. The other two pipe symbols are marked *6" S.S.,* meaning a 6" inside diameter sanitary sewer line. The black triangle tips of the symbols show the direction the line is flowing.

Sewer service lines off the main line are indicated with a small triangle symbol with an *S* in it. The circles drawn in the main lines represent manholes. Notice the 10" S.D. ends at the intersection of Zenith Drive and Briartree Way. From the manhole at that intersection, four lines are shown radiating out from the manhole. These end at four small black rectangles that stand for gutter drain inlets.

The water system is not shown on this plan. Usually the water system and the electrical street lighting layout are on a separate sheet.

Highway Project Plans

The plan drawing for a highway is similar to the plan drawing of a subdivision. Look at Figures 2-5 and 2-6, then turn back to Figures 2-3 and 2-4 and compare the two sets of drawings. Notice that they are very similar. The highway profile, Figure 2-6, is very simply drawn. Notice the elevations from 16 to 18 on the left edge and the station numbers from 195 + 00 to 200 + 00 along the bottom

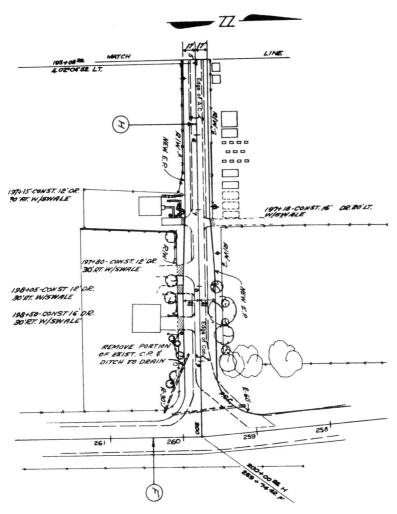

Highway plan sheet
Figure 2-5

edge. Each rectangular segment of the graph paper used for this profile represents 0.20 (two tenths of a foot) vertically and 50 feet horizontally. Check the rectangular segments with the station numbers and elevations given and you will see that this is so. The *N* at the left of Figure 2-5 shows north.

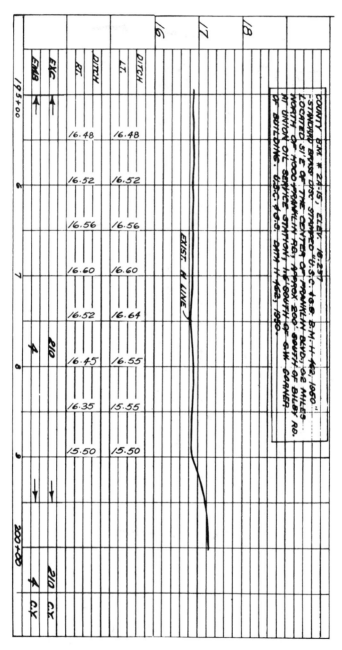

Highway profile sheet
Figure 2-6

On Figure 2-6, the irregular horizontal line through the middle of the graph paper is called the *existing H line*. This is a reference line used in drawing the plan and profile. The architect in this case designated the centerline of the road as the "H line" rather than the "centerline." This profile doesn't show the elevation of the finished H line. Usually the highway profile drawing has more details and would be more like the profile drawing for the subdivision, Figure 2-3.

A finished H line is not shown on the profile drawing because, except for the asphalt overlay, all the work is to be done adjacent to the existing pavement. Notice Figure 2-7 (cross section). It shows all the ditch and road section excavating adjacent to both sides of the existing road. The only time the existing H line shown in Figure 2-6 will change is when the shoulder widening is complete and the final asphalt pass is made over the existing road and H line.

Look at the cross sections in Figure 2-7. Notice the existing H line elevations written vertically at each station. The second elevation written on an angle and underlined is the finished H line elevation. Notice that the largest grade change at any station between existing H line and finished H line is 0.20 foot. This profile drawing also does not show any underground pipe, except the short piece to be removed on the right side at station 199 + 50.

Notice the left side of the highway cross section says "Lt" and the right side says "Rt." *Rt* means right and *Lt* means left. The ditch elevations for both sides of the road are given at 50-foot intervals. The right side of the road is the side that would be on your right if you stood with the smallest station numbers to your back and looked down the road toward the larger station numbers. In Figure 2-5, the right side of the road would be the south side of the road and the left side would be the north side.

At the bottom left of the profile, Figure 2-6, are two abbreviations: *Exc* and *Emb*. These stand for *excavation* and *embankment*. Look across the sheet to the figures *210 C.Y.* and *4 C.Y.* written at the lower right. This means that there are an estimated 210 cubic yards of excavation and 4 cubic yards of embankment in this section of roadway. Some highway profile drawings have the

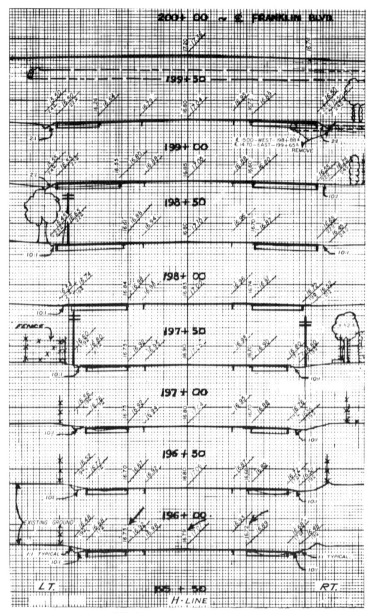

Highway cross section
Figure 2-7

estimated yardage figured to help the foreman and estimator. A glance at each page of the profile will show you where the major cuts and fills will be made. A foreman looking at the figures in Figure 2-6 would see that there will be 206 cubic yards of excess dirt to haul off or move to an embankment further down the road.

Every job is staked with a point of known elevation called a *bench mark*. Read the information in the rectangular box at the top of Figure 2-6. It explains how to locate the bench mark used by the surveyors to establish the grade for the project. The elevation given, 16.237, will be rounded off to 16.24, or 16 feet and 24 hundredths above sea level.

Refer again to Figure 2-5. The station numbers on the H line correspond with the station numbers in the profile drawing. The width of each half of the street is shown: 17 feet at station 5 and 22 feet at station 8. You should understand that station 8 is more fully described as station 198 + 00.

Notice that Figure 2-5 shows the location of all the trees, fences, houses, and two sections of pipe at the intersection. The width of the existing road is pointed out in two places with the words *Edge of A.C.* (asphaltic concrete) and *Edge of Conc.* (concrete). The writing at the left and right of Figure 2-5 gives the details of where the driveways must be built. The driveway instructions at the right of the sheet give the station number for the center of the driveway, 197 + 18. It then indicates that a 16-foot wide driveway must be built at station 197 + 18. The driveway will extend from the edge of the new road to 20 feet left of the H line. A *swale* is a shallow dip in the driveway to let the ditch water pass through. A swale is used rather than placing a pipe through the driveway to let the water pass under.

Cross Section Drawings

Every plan and profile drawing such as Figure 2-5 and 2-6 must have a cross section drawing giving you the information you need to build a highway. Typical cross section drawings are shown in Figure 2-7. Notice that the station numbers in Figure 2-7 are the same as the station numbers in Figures 2-5 and 2-6. Figure 2-7 is a cross section view of the highway in Figure 2-5 at every 50 foot station. The smallest station numbers are at the bottom on the cross section drawing. Visualize this as if you were standing in the road with the small stations at your back. This puts the *Rt* ditch on the right side of the sheet and the *Lt* ditch on the left side of the sheet.

The cross section drawing is on graph paper. The heavy black lines are at 10 foot intervals. Each small square represents a 6'' square. Several numbers are written diagonally above each cross section. These are finished road and ditch elevations. Below the two diagonal lines at the left and right sides of each cross section, there is a number. This is the distance in feet to the edge of the pavement and to the ditch from the centerline. The distance on the right and left from station 195 + 50 to 197 + 00 is 17 feet to the edge of the pavement and 19 feet to the ditch. From station 197 + 00 to 199 + 00, the ditch and edge of the pavement widen or move further from the centerline and reach 22 feet to the edge of the pavement and 24 feet to the ditch at station 199 + 00. These distances are measured from the centerline, or as in this case, the H line.

Numbers are written vertically and pointed out by arrows above station number 195 + 50. These are the elevations of the existing road surface at the centerline and at each edge of the existing pavement. Study the cross section drawing. Notice that full structure sections are to be built on each side of the existing road. Follow the dark irregular line running horizontally through the drawing. This line is the existing ground elevation before the ditch and road sections are built. The existing ground line is pointed out in the drawing at station 195 + 50. The rate of slope for the front and back of the ditch is indicated as *1:1 typical* and *10:1* respectively at station 195 + 50.

At station 199 + 00, on the right side of Figure 2-7, there's information for removing drain pipe. The flow line elevations and station numbers of each end of the pipe are given. Subtracting the small station number from the larger number gives you the length of pipe to be left, in this case 77 feet.

At the right and left of each drawing, several trees and fences are drawn in. This is done to insure that the contractor leaves them in place, and to show the clearance between these objects and the ditch slope. In the ditch areas on each side of the drawing at station 197 + 00, a power pole is shown. At station 198 + 00, another power pole is indicated in the section to be built. That pole must be moved before or during construction. In some cases you'll have to work around a pole.

Along with the plan drawing, profile drawing and cross section drawing, Figures 2-5 through 2-7, the highway plans will include a typical road section sheet similar to Figure 2-1. Always study these plans carefully before beginning a project.

You can make difficult plans easier to read if you color code them. For example, all the new widening areas could be colored green, the overlay areas red, and the pipe runs yellow. This would help you distinguish each excavation area more easily.

Occasionally you find a symbol or notation that doesn't make sense to you or that you haven't seen before. Even experienced excavation contractors may come across something on a plan that's not clear. In every case, the surveyor or engineer who developed the plans should be able to answer your question.

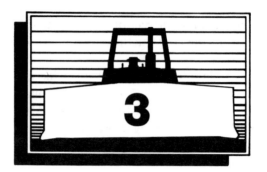

Using a
Contour Plan

Most of the plans in the last chapter used cross section views to indicate changes in the slope of the ground. For example, Figure 2-3 shows the existing ground and finished grade in profile. For many jobs, all you'll need is a cross section view of the finished grade. But on some jobs you'll be given a more exact description of what finished grades should be when work is completed. A contour plan (sometimes called a topographic plan) provides that information.

Reading a Contour Plan

Figure 3-1 is a contour plan of a landscaped area in front of an industrial building that's being constructed on Round Hill Lane. It shows finished elevations with contour lines. Each contour line on the plan connects points of equal elevation. Where the lines are closer together, the ground is to be steeper. Where the lines are farther apart, the slope is very gradual, or even flat. See if you can

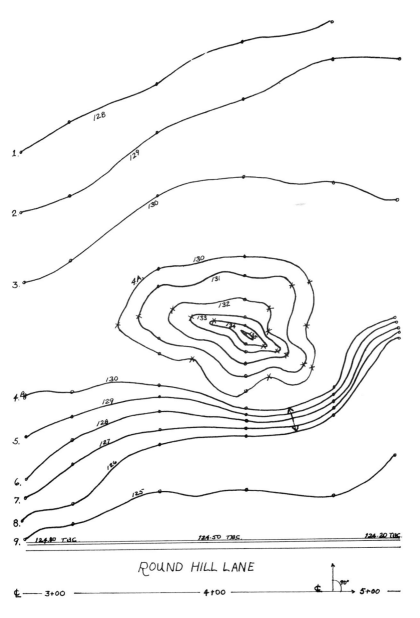

Typical contour drawing
Figure 3-1

visualize the ground this plan represents before we start discussing
it.

The Contour Interval
On this plan each contour line shows a rise or fall of one foot of
elevation. Some plans have a contour interval of 5 feet or more.
For our convenience, I've numbered these contour lines down the
left side of the diagram. Look carefully and you'll see that there
are two lines numbered 4. Both represent an elevation of 130 feet.
Line 4B extends from the left edge of the figure to the right. Line
4A is a closed loop, encircling the mound at the center of the
figure.

Start at the top of the figure and follow the first four contour
lines. You'll see elevations of 128, 129, 130 and again, 130. The
first three lines are spaced further apart, showing a gradual rise
toward the center of the figure. The elevation numbers go from
128 to 130, indicating a 2 foot rise in elevation. Remember, the
larger the number, the higher the elevation.

Now look at lines 3, 4A, and 4B. All three of these contour lines
are at an elevation of 130, indicating that this center area is flat —
except where the five contour lines form circles inside line 4A.
These lines, numbered 130 to 135, show a rise of 5 feet. From this
figure you should be able to picture the shape of the ground when
excavation is complete: an irregular shaped mound 5 feet high.
Notice that the contour lines that indicate the mound get closer
together at the bottom right of the mound. This means the slope is
steeper at that area.

Closed Loop Contour Lines
Contour lines that form a closed loop always indicate a mound or
a depression. If the closed circular contour lines went from elevation 130 to 125, instead of from 130 to 135, you would be looking
at a 5-foot depression rather than a 5-foot mound.

Now look at the group of contour lines from 4B to 8. Following
these lines to the right, they move closer together. This shows a
slope that becomes steeper at the right of the figure, dropping 4

feet from the top contour line of 130 to the bottom contour line 126. If this plan were drawn to scale, you could find the slope at the arrow by measuring the length of the arrow. If the arrow represented a length on the ground of 8 feet, then we would know that the slope was 2:1 (8 feet of horizontal distance in 4 feet of vertical distance). In any contour plan drawn to scale, you can measure a distance on the plan and compute the slope.

Between contour line 8 and contour line 9 there's only a 1-foot difference in elevation. You can see that the ground at the bottom of the slope is nearly flat, especially at the right of the figure. At the bottom of the figure you'll see elevations marked 124.80 T.B.C., 124.50 T.B.C., 124.20 T.B.C. These are the elevations at the top back of the curb on Round Hill Lane.

Now check contour line 9. It's at elevation 125. Since all the top back curb elevations are lower, the ground must slope gradually from contour line 9 to the top back of the curb.

On Round Hill Lane, the figure shows the centerline and the stations, numbered every 100 feet, starting at 3 + 00 and going to 5 + 00. Using a scale drawing like Figure 3-1, a grade setter and foreman could set stakes in the contour area for grading.

Staking the Area

Here's how to set out grade stakes for the excavation work that's going to be done on this lot. Start by setting up a surveyor's transit over the centerline mark at station 5 + 00. Sight up the centerline to station 4 + 00. This gives us an exact reference line for the other sightings we'll take. Set the transit at zero degrees along the centerline and turn 90 degrees toward the lot, as indicated on Figure 3-1. Now scale from the plan the distance to each of the contour lines directly out along the 90 degree line. Set a stake in the ground at each contour line, sighting through the transit for line. Continue this procedure every 50 feet down the centerline. Drive a hub in the ground at each stake as a reference for shooting elevations.

When stakes have been set as indicated by the small circles drawn on contour lines in Figure 3-1, you may want to set intermediate stakes where contour lines make abrupt changes of direction. These stakes are indicated by the X-marks made through contour lines in Figure 3-1. Set these intermediate stakes by measuring from stakes set along the 90-degree lines. Intermediate stakes will give better control once the grading has started.

Marking for Cut or Fill

The next step is to mark the amount of cut or fill needed at each stake to produce the finished grades shown in Figure 3-1. We have to start from some known elevation on site. Use a bench mark if one is available. On this site, we don't have a bench mark, so we'll use one of the top back of curb elevations as our starting point.

Hold a surveyor's rod vertical on one of the top back of curb marks and sight level from the transit to the rod. If you're using a laser, slide the receiver unit until you hear a steady signal. If the top back of curb elevation is 124.50, loosen the clip that holds the rod tape and move the tape to read 4.50 at the point the transit is sighting to the rod, or the laser receiver indicator has a steady signal. Once a 4.50 elevation is in place and level with the transit sighting or laser beam, lock the rod tape by screwing the clip tight again.

Now we can start marking cuts and fills at each of the hubs. Set the rod vertical on the first hub. Without disturbing the transit or laser unit, sight on the rod or receive a steady signal from the laser receiver. Suppose the rod reading is 6.50 feet with a level sighting or signal. Only the last three numbers of the elevation can be set on the rod, so the 6.50 is actually 126.50. If the contour plan shows that the finished grade should be 128.00 at that point, we would mark a fill of 1.50 feet on the stake by that hub. We calculate the fill by subtracting the 126.50 actual elevation from the 128.00 shown on the contour line.

Continue sighting on the rod and marking cuts and fills until the work to be done has been noted on every stake.

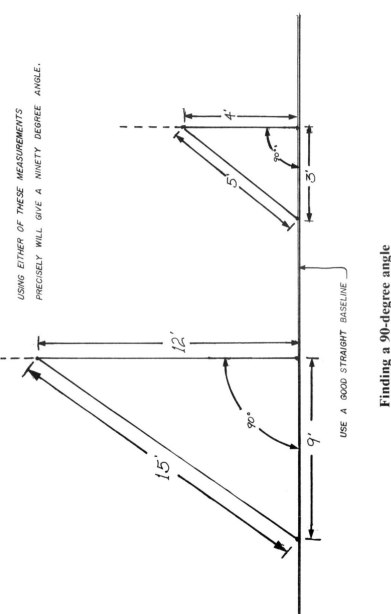

Finding a 90-degree angle
Figure 3-2

Making a 90-Degree Angle

If no surveyor's transit is available, you can still set out the stakes shown in Figure 3-1 using just a tape measure. First, find the centerline of the road. Make a 90-degree angle from that centerline by constructing a 3, 4, 5 right triangle. See Figure 3-2. Measure 9 feet along the centerline. Then measure 12 feet out from the baseline. Now connect the ends of two lines. Adjust the perpendicular line until the line connecting them is exactly 15 feet. When you're at 15 feet, the angle is 90 degrees. Then set out the stakes along the perpendicular by measuring on the contour plan.

When all the stakes set along the 90-degree line from the centerline have been marked at each contour line, and the intermediate stakes are placed, the grading can begin.

Most contour plans show the existing contours in dashed lines and the contours to be built in solid lines. Don't be confused when you see the extra lines on a plan. We left the existing ground contour lines off the drawing in Figure 3-1 so it would be easier to read. But the contour plan you get may show both existing and finished grades.

Generally the accuracy of finish grading for apartment landscaping or industrial tract landscaping isn't that critical. Since extreme accuracy isn't usually important, the excavation contractor rather than a surveyor will often set his own grades. A hand-held sighting level will usually be good enough for setting grades in that case.

If you look at all the circles and X's on Figure 3-1, you'll probably agree that heavy excavation equipment wouldn't have room to work between them. The initial staking will give the operators an idea of what's needed. Then the operators should work around the stakes as long as possible. Eventually the equipment will have to work over the stakes. For final grading, a few of the obliterated stakes may have to be reset from other stakes that weren't run over by equipment.

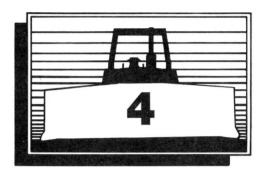

Grade Setting

Setting grade takes skill, knowledge and attention to detail. The grade setter transfers information from the plans and surveyor's stakes to stakes that the excavation equipment operators read and follow. The grade setter must be both fast and accurate; fast to stay ahead of the equipment and accurate because any error can be very costly. In many cases, the foreman will not have time to check the grade setter's work and an error may not be discovered until days later — when it becomes very obvious to the foreman and grade setter alike.

A good grade setter must be able to perform rapid mental arithmetic. His entire day is spent adding, subtracting, and multiplying as he transfers grades and distances off the original surveyor's stakes and sets his own stakes and grades. He must set his stakes at every key grade change. And they must be located where they're not in the way of the equipment but where the operator can see them. A grade setter must stake the tops of slopes to be cut, the slopes themselves as they are trimmed to grade, and

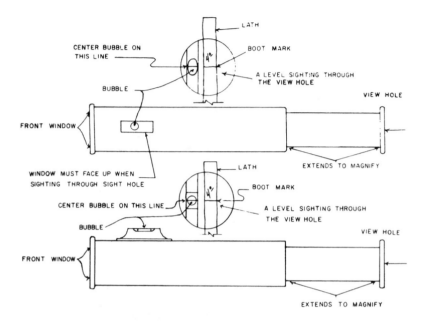

Two types of eye levels
Figure 4-1

the toe of the slope once it is reached. He must set stakes at ditch bottoms, road shoulders, road centerlines, and so on.

Keep the Eye Level Accurate

Grade setters use a small hand-held sighting level to transfer grades from one point to another. See Figure 4-1. Because this instrument gets rough use, it should be checked for accuracy several times a week. Here's how to do it.

Set up two 4-foot high stakes 30 feet apart. Wood lath makes the best stakes for this purpose. Clip a clothespin on each lath. Stand beside one lath and sight on the other. Adjust the level of the clothespin on the lath near you until it's at the same level as the clothespin on the lath 30 feet away. Then, reverse the process. Go to the distant lath and sight back on the first lath. Sighting back, if both pins are at the same level, the instrument needs no adjustment. If you sight over or under the clothespin you first sighted from, then the eye level is off and needs adjusting.

To adjust the eye level, unscrew the front lens and turn the front screw one turn. Repeat the test described in the paragraph above. If the level is off by more than in the first test, you adjusted the screw the wrong way. Turn the screw back to the original point, and then one more turn. Keep adjusting the screw and checking the two laths until you're sighting level in both directions. If you can't bring the level into adjustment, return it to your dealer for repair. Even the most experienced grade setter can't do a good job with an inaccurate eye level.

Crows Feet

A grade setter usually won't drive a hub into the ground to mark grade levels. Hubs are needed only when it will be a permanent point where many elevation shots will be taken or when trimming to a close tolerance. Instead, he'll indicate cuts and fills for equipment operators with markings on a wood lath. Most grade setters call a lath like this a *crows foot*. See the laths marked (1), (2), and (3) in Figure 4-2.

(1) The circle through the tail of the arrow always means that the point of the arrow is drawn to the finish grade. The operator seeing this lath knows that is the elevation he must grade to.

(2) This crows foot does not have the circle drawn through the tail of the arrow. Instead there's a *C* followed by a number. This means that more cutting is needed. One-half a foot (0.5 feet) must still be excavated below the line marked across the lath.

(3) This crows foot or lath is similar to (2), but it would be used on a fill slope. It gives both the fill still needed at the lath (up to the horizontal line) and the fill needed 10 feet from the lath (5 feet). The 2:1 slope was computed by dividing the 5 feet remaining to fill into the 10-foot distance from the stake.

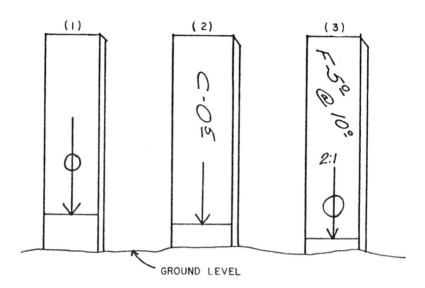

Crows feet set by grade setter
Figure 4-2

Setting up a line of laths and marking them is usually called *setting crows feet.* This method is faster than setting hubs because the top of the stake is not set on grade. On most jobs, the equipment will eventually be directed to make a pass right over the crows foot lath line. Then the lath will have to be reset.

Where an extremely accurate grade must be cut, the grade setter will replace his laths with hubs once the grade is nearly established. The hub marks a point on the ground like a lath but it also establishes a precise elevation at the top edge. After setting the hubs, the grade setter should drive a short lath next to each hub so it can be seen by the equipment operators.

Setting a Crows Foot Grade or Hub

To set a crows foot grade or hub, the grade setter must establish the ground elevation at a given point in relation to the surveyor's hub. Assume the surveyor's information stake reads *C-3[50] centerline 35[0]*. The grade setter must measure the 35 feet from the hub or reference stake, whichever is indicated. There he drives a lath or hub.

Then the grade setter sets his ruler (with small numbers down) on the hub or at the base of the lath. If the hub is used it should be only partly driven in the ground at this time. Then he adds in his head the road section thickness to the cut computed by the surveyors. Assume the road section is 0.50 foot. He'll add the 0.50 foot to the C-3^{50} to get C-4^{00}.

Next the grade setter shoots across to the surveyor's hub or the RS point until his eye level reads level. To get a level shot, the bubble must be centered and his eye level line and top of hub or boot line level with each other. When he has a level shot, he looks to see where the eye level is in relation to his ruler. If the eye level is at 3.8 feet on the ruler, he'll drive the hub 0.20 foot lower so the level line intersects the ruler at the 4-foot mark. If a crows foot lath is used, the grade setter draws a horizontal line at ground level with an arrow pointing to the line. Above the arrow he would mark C-0^{20} so the equipment operators will know a 0.20 foot cut is needed at that point.

Setting Boots

In many instances, obstruction will obscure the line of sight to the hub or RS. Then the grade setter will set up *boots* before any work begins. To set a boot, the grade setter drives a 4-foot lath behind the hub the surveyors have set. Then he rounds the surveyor's cuts off to an even foot. The C-3^{50} the surveyors set would be rounded up to 4 feet. The grade setter measures from the hub up 0.50 foot and draws a horizontal line on the 4-foot lath at that point.

If he needs to measure up the lath further to clear any obstacles in his line of sight, he may raise the boot another foot or more. Assume that 0.50 foot is enough to clear all obstacles. Then there will be a cut of 4 feet to the finished grade from the horizontal line marked on the 4-foot lath.

The grade setter rounds all the cuts off to 4 feet and marks all the lath up that distance from the surveyor's hub. For example: C-2^{60} plus 1.40 equals 4 feet. C-3^{10} plus 0.90 equals 4 feet.

If a fill rather than a cut is marked, he'll measure up the amount of fill marked by the surveyors plus 4 feet. For example, if the surveyors have marked F-0^{10}, he'll measure up 4.10 feet to equal a 4-foot boot. For F-0^{40}, measuring up 4.40 feet equals a 4-foot boot. Doing this at every hub makes the grade setter's job easier and faster. Then adding the 0.50 foot road thickness to the ruler reading at each station will give a 4.50 cut at every station that he sets.

When the boot needed is higher than a single 4-foot lath, he can drop to a 3-foot boot or nail two lath together for more height.

Sometimes the grade setter can't set crows feet because there's no room for them. He must then check the grade after each scraper or grader pass and let the operator know how much more cut or fill is needed.

Staking for a Typical Cut and Fill Station

Refer to Figure 4-3. The only hubs set are those set by the surveyors. The cuts, fills and distances were computed from these hubs. First, the surveyors established the centerline and staked it as shown. The existing ground elevation at the edge of the shoulder was established from these stakes. The surveyors then determined the cut or fill needed at that point from the elevations shown on the job plans. The amount of cut or fill must be known so the surveyors can compute the distance needed for slopes and ditches from the edge of the shoulder to the point where the hub must be set on each side of the centerline.

Then the surveyors computed the distances needed and drove the hubs required on each side of the centerline at each key station. Then they shot the elevation at every hub. Once the elevation of every hub was established, the cuts, fills and centerline were computed and marked on the information stake next to the hub. At this point the surveyors have supplied the contractor with all the information needed to build the road.

Look again at Figure 4-3. Except for the R/W (right-of-way) stakes, all stakes have been set by the grade setter and all are crows feet. No hubs have been set by the grade setter. Notice the two crows feet marked *Sho* (shoulder) and *HP* (hinge point). Some

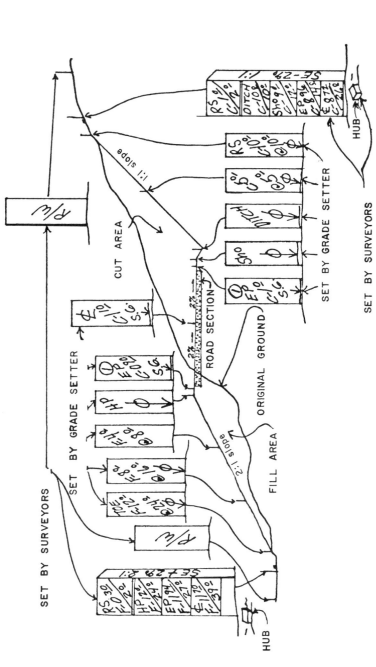

Staking for typical cut and fill section
Figure 4-3

grade setters set hubs at those points if the grader operator doing the trimming needs or requests them for more control and a closer cut. The shoulder and hinge point grade will be fine trimmed after the road has been paved and should be left a little high at this time; two tenths of a foot (0.20) should be about right.

Once the dotted area in Figure 4-3 labeled *road section* has been cut out and compacted, the crows feet at the edge of the pavement and at the centerline will be replaced with hubs so the fine trimming can be done accurately.

Location of Stakes

The grade setter should give some thought to the best place to locate the stakes. Timing can also be just as important as the actual location. No matter where or when the stakes are set, keep the stakes or hubs set on firm ground or aggregate. For example, suppose hubs are needed in an area where fill has been brought up to the grade marked on laths. The grade setter must be sure to pull the stakes or offset them so the area where the stakes have been can be compacted before setting the hubs. In a deep fill area, the lath must be offset periodically so the area where the grade laths are placed can be filled and compacted. Then new lath is set back with grades for the remaining fill.

Never set stakes so close to the area being filled that the edge of a fill can't be compacted without covering the stakes. All fill hinge points must be over-filled so that, when trimmed, the hinge point will be well compacted. If the shoulder is to be rocked or paved and the subgrade at a hinge point was not over-filled slightly, the rock or pavement will be lost over the edge of the slope. See Figure 4-4.

If a dike will be placed on the edge of the pavement, the grade setter or foreman must check the edge of pavement (EP) distance on the surveyor's stake. The surveyor's stake may give the distance to the front or back of the dike. There must be 3 or 4 inches of asphalt concrete behind the dike so the dike machine can function properly. The grade setter should adjust his stakes to give enough of an offset to allow for the extra 3 or 4 inches needed to place the dike.

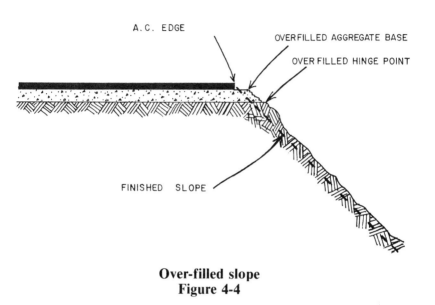

A.C. EDGE

OVERFILLED AGGREGATE BASE

OVERFILLED HINGE POINT

FINISHED SLOPE

Over-filled slope
Figure 4-4

The grade setter must be very careful when setting lines and grades on all edges. Top of slopes, toe of slopes, hinge points, choker lines, and ditch lines must be set precisely, for line as well as grade.

The grade setter should never assume the correct cut or fill was made by the equipment operator. After the operator has made the cut or fill, the grade setter should recheck the grade. Both the grade setter and the foreman should watch closely as fills and cuts are made. It's easy for an operator to overlook a less accessible area where cuts and fills are harder to make. Check these areas carefully before the equipment moves to a new part of the job. Moving the equipment back later wastes time and effort.

It's up to the grade setter to see that no small areas or edges are left half finished. He should also call the foreman's attention to any equipment operators who run over stakes unnecessarily. The grade setter is busy enough without having to replace stakes that have been carelessly run into the ground.

Grade Setting Equipment

The grade setter needs the right equipment to be efficient at his job. He should have a pouch to carry a 50 foot or 100 foot tape, a marking pencil or crayon, ribbon or red spray paint, and a ruler. He should have a hatchet and an eye level on his belt. The grade setter also needs a sack, bucket, or carrying rack to carry stakes and hubs. If working in hard ground, a steel pin may be needed to drive a hole for the wood hubs. He needs some fluorescent red paint to spray the hub tops and lath tops so they can be easily seen. Spray painting is much faster than tying colored ribbon on the laths or marking blue crayon on hubs. Spray-painted hubs are much easier to find after the grader has passed over them.

Using clothespins to set grades— Clothespins can be used to indicate cuts and fills. This is usually the most efficient method for three reasons. First, it eliminates one man, the guinea hopper, who would otherwise be required to point out the hubs to the grader operator. Second, it saves much of the time normally required to set grade hubs. Finally, it saves the cost of extra hubs and laths. Clothespin grade setting is the quickest and least expensive method to set grades when there are curbs on both sides of the road.

The grade operator must trim one side along the curb to grade. The grade setter then drives a 4-foot lath against the edge of the curb at 50-foot intervals. After driving the lath, the grade setter marks a horizontal line on the lath, measuring up 3 feet above the top lip of the curb. A colored survey ribbon or crayon makes a good horizontal mark. Figure 4-5 shows a street using only clothespin grade stakes without centerline hubs. The clothespin shows there is a slight cut remaining at the centerline. When the centerline is on grade, the clothespin will be set on the ribbon.

If there are no curbs, set the 4-foot lath 1 foot behind the point to be trimmed at the edge of the road and opposite each grade station. For example, if the surveyor's information stake gives a distance of 20 feet to the edge of the pavement, measure out 19 feet and drive the 4-foot lath. Next check the elevation given for the edge of the pavement. Mark the horizontal line on the 4-foot

**Clothespin grade stakes without
centerline hubs
Figure 4-5**

lath 3 feet above the level of the finish edge of the pavement.
Follow the same steps at every grade station. Then use the procedure explained in the previous paragraph for setting the grades.

Once all lath are placed and the horizontal lines or ribbons are
set, the grade can be checked. If the road has a crown in it, the
grade setter must subtract the amount of crown from the 3-foot
horizontal line and add the thickness of the road section. This
gives the rod reading desired at the centerline. For example,
assume that there is a 0.30 crown. Subtract that from the 3-foot
mark on the lath and you get 2.70 feet. Add 0.50 for the thickness
of the road and you get 3.20 feet. The 3.20 is the height the grade
setter should sight with his eye level at each station to be on grade
if all the stations have a 3-foot boot.

Using an eye level to check grade— The grade setter measures out from the lath to the centerline. Then he sights back with his eye level along a level line to the horizontal line on the lath. This is done by standing the ruler up with the zero mark on the ground. Slide the eye level along the ruler until it sights level. Once it sights level, read the numbers on the ruler at that point.

Let's assume that at that point the ruler reading is 3 feet. To be on grade, the ruler should read 3 (feet) and 0.20 (two tenths). This indicates 0.20 (two tenths) must be cut from centerline before the correct grade is reached. The grade setter will now clip a clothespin on the 4-foot lath 0.20 (two tenths) above the horizontal mark (which is the 3-foot boot). Then he moves to the next lath and follows the same procedure. This time he reads 3 (feet) and 0.70 (seven tenths) on his ruler. Then he clips a clothespin on the second lath 0.50 (five tenths) below the horizontal mark. This means a 0.50 foot fill is needed. The grade setter will continue shooting each lath, clipping a clothespin above the horizontal line for cuts and under the horizontal line for fills. If the ruler reading is less than 3.20, a cut must be made. A ruler reading over 3.20 means that a fill is needed.

The grader operator can see the lath and knows whether a cut or fill is needed by where the clothespin is in relation to the horizontal mark on the lath. The grade setter need not stay with the grader operator. When the grade setter finishes marking the last station, he should double back and check the work the grader operator has completed. The grade setter will adjust the clothespin after reshooting the grade. Once the grader operator finishes the first pass, a second pass can be made starting from the beginning. The grade setter will have adjusted the clothespins, placing them the exact distance above or below the horizontal line as needed for a cut or fill.

When the centerline is on grade, the clothespin is placed on the horizontal line. When the grader operator sees the clothespin clipped to the horizontal mark, he knows that no further work is needed at that station. Survey ribbon tied around the lath in place of horizontal crayon marks works well because it can be pulled off when finished. The lath can then be used again.

If the road is wide and the grader operator has trouble determining exactly where the centerline is because he has no centerline hubs to follow, the grade setter should mark the centerline with spray paint each time he takes a grade check. This method of checking grades can also be used for roads that don't have curbs.

Centerline and shoulder grade can both be set on the same lath by using two clothespins painted different colors. One color can be used for the centerline grade and another color for the shoulder grade.

String Lines

There are many types of excavation work where the grade is set best with a string line. These include laying pipe, setting forms or preparing for a curb machine. The string line should be set with great care and stretched as tight as possible. A good nylon line is best for this type of work. In most cases, pipe laying requires a string line. Before setting the line, determine how much of an offset is needed so the string will not interfere with the equipment to be used. If a trencher is used, the string line must be on the operator's side — the side opposite where the spoil is thrown. If a backhoe is used, be sure to keep the string line back far enough to clear the equipment outriggers.

When setting string line, keep three stations set up along the line so you can sight down the line to correct errors or movement which might occur during setting or excavation. A faulty station grade will be easily detected this way. When you sight down the string line, each station should blend with the next with no sudden rise or dip from one station to the next. If you detect a sudden grade change, check the measurements. Check the rate of slope shown on the plans if you can't find any error in your work. If the plans do not show a sudden grade change, the surveyors have made the error and it should be called to their attention.

If waiting for the surveyors to correct a faulty station would hold up production, consider correcting it yourself. Set up three more stations with the string line and sight through the obviously faulty station, lowering or raising the string until it flows smoothly into the correct grade. Usually the surveyors will have made an error at only one station and this will solve the problem without holding up the trenching operation. Always be sure the string line is straight. Look for any slight variations. The trencher, hoe, or automatic machines will use the line for direction as well as elevation.

If the steel pins used to hold the string line are offset to one side from the grade hub, use a carpenter's level and straightedge to transfer your grade from the hub to the grade pin. An eye level may be used, but this is not good practice where a close tolerance is required. If the distance is too far for a straightedge, use a surveyor's level or laser level.

Setting Offset Strings

Assume now that you are ready to transfer the surveyor's grades to the grade stakes. The surveyors have set stakes for 500 feet of 30-inch drain pipe. The contractor should have requested the distance the surveyor's stakes are offset from the actual trench location and the side of the offset. The width of the trench and direction the spoil will be thrown determines these two factors.

Let's assume the surveyors have set all their grade hubs on a 10-foot offset. A trencher is to be used for the excavating. The distance the string is offset from the surveyors' hubs is determined by the type of trencher. Measure from the center of the bucket line to the grade indicator arm. This arm extends out along the bucket line and can be adjusted in height. A metal wand projects horizontally from the end of the grade indicator arm. Once the required depth is reached, the grade indicator arm is lowered until the wand is touching the string. It will be kept in that position by the operator by raising or lowering the bucket line. See Figure 4-6. A second rod extending from the engine end of the trencher is used to keep the direction of travel straight.

If the measurement from the center of the bucket line to the end of the wand is 8 feet, the string line should be set 2 feet from the surveyor's hub. With the hubs at a 10-foot offset, this will put the string line 8 feet from the center of the trench.

**Wand held just above string
for grade and at mark for line
Figure 4-6**

Setting String Line Height

For a job like this there should be no less than four grade pins set at 50-foot intervals along 150 feet of trench at any time. Once the string line pins are up, the grade setter must transfer the grade given by the surveyors to the grade pins at an even foot. Assume that the terrain is reasonably level and the first four grades on the surveyor's stakes read as follows: C-6^{80}, C-6^{13}, C-6^{03} and C-5^{90}.

The string line would be set for these cuts as follows. An 8-foot high string line should be used because a 7-foot string would only be 0.20 foot off the ground at the first grade. A 9-foot line would be more than 3 feet above the ground at the last grade. A line 8 feet above the flow line of the pipe is high enough to clear most ground obstructions yet low enough to step over easily. To set an 8-foot

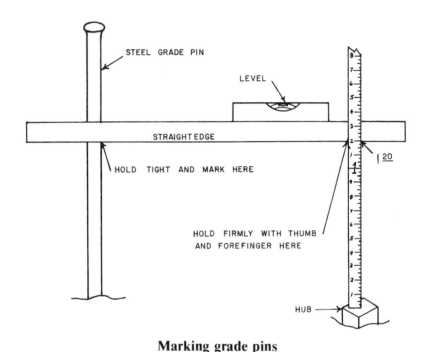

STEEL GRADE PIN

LEVEL

STRAIGHT EDGE

HOLD TIGHT AND MARK HERE

HOLD FIRMLY WITH THUMB
AND FOREFINGER HERE

HUB

| 20

Marking grade pins
Figure 4-7

string line, add to each cut the number of hundredths and feet
needed to total 8 feet. For example:

$$
\left.
\begin{array}{l}
\text{C-6.80} + 1.20 = 8 \\
\text{C-6.13} + 1.87 = 8 \\
\text{C-6.03} + 1.97 = 9
\end{array}
\right\}
\begin{array}{l}
\text{Even foot} \\
\text{eliminates hundredths}
\end{array}
$$

You can get the same answers by subtracting the cuts required
from the 8-foot string line. Use the method which is easiest for
you. For the 6.80 cut, measure up the grade pin 1.20 feet.

Set a ruler on the surveyor's hub as shown in Figure 4-7 and
measure up 1.20 feet on the ruler. Now set the straightedge against
the 1.20 mark on the ruler. Hold the straightedge firmly against
the ruler with the thumb and forefinger. Set a torpedo level on the
straightedge and level the straightedge, holding it against the grade

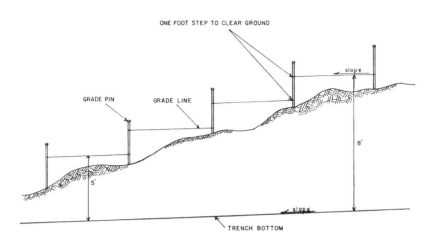

ONE FOOT STEP TO CLEAR GROUND

slope

GRADE PIN   GRADE LINE

8'

5'

slope

TRENCH BOTTOM

Stepped grade line
Figure 4-8

pin. Once the straightedge is level, have someone mark the grade pin. Be sure that if the bottom of the straightedge is held at 1.20 feet that the pin is marked from the bottom of the straightedge and not the top. Tie the string at the mark on the grade pin. Follow the same procedure at each station to get a string line exactly 8 feet above the flow line.

If the terrain is steep, the grade line may have to be changed often to keep the string line from being too high or running into the ground. If there are large grade changes, the string must be tied in two spots on each grade pin and two measurements must be figured. See Figure 4-8.

Direct Overhead String Lines

Some inspectors don't think that an offset string line is accurate enough for checking final hand grading and pipe laying. Direct overhead line may be required. If this is the case, the grade line is set in exactly the same way, the only difference is that *T-bars* are used instead of grade pins. See Figure 4-9. Instead of tying the string around the T-bar upright where the mark was made, a nail is driven in the board at the mark. Drive the nail in just far enough to

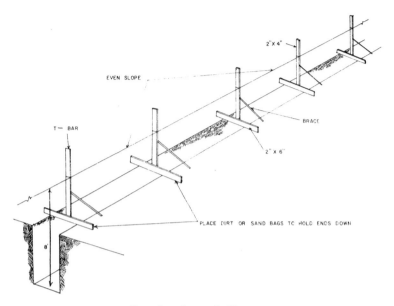

Overhead grade line
Figure 4-9

make sure it cannot be pulled out by hand. The ends of the T's must be weighed down with earth before the string is stretched. Brace the T-bars with a 1 x 2 board so they won't be pulled over when the string is stretched.

After all the bars are marked and nails driven at the marks, tie the string line. The string line can be tied directly to the nail on the first board, with one wrap around the bar first. On the next bar, the string is pulled under the nail and wrapped around the bar once and back under the nail. Now pull the string tight from that bar to the next. When the string is stretched tight, flip it from beneath the nail to the top side. This will lap it over the bottom string and hold it from slipping. Note Figure 4-10. The same method is used on grade pins, except the string is looped twice around a grade pin.

Setting a Bottom String Line

Some foremen prefer a string line at the bottom of the trench so the grading person can check his own grade. To set a bottom line,

BOARD LOOP

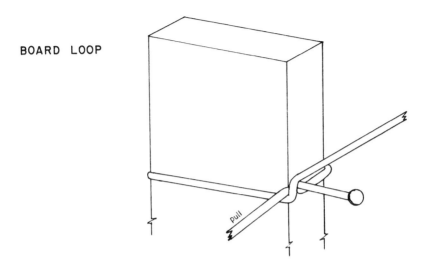

PIN LOOP

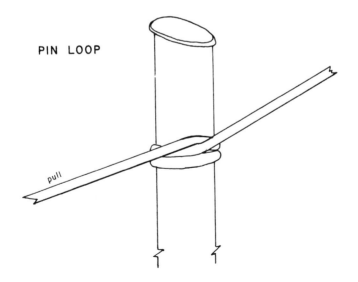

**Tying the string line on
T-bars and grade pins
Figure 4-10**

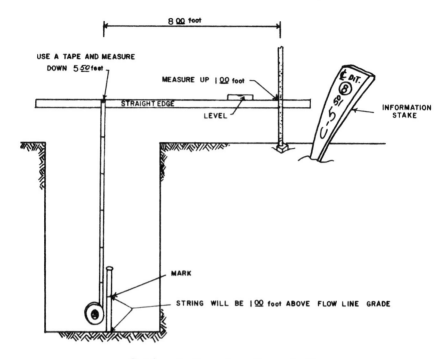

Setting bottom trench grade line
Figure 4-11

the grade pins are driven at the bottom of the trench below the stations set by the surveyors. A straightedge is used as if you were setting the top line. Measure up from the grade hub 1 foot and on a level line across to the trench. Now measure down with a tape measure to the cut indicated on the surveyor's lath and mark the pin. See Figure 4-11. Do this at each station. When the string is set on those marks it will be 1 foot above the flow line of the pipe. Three people are needed to set a bottom line. One measures at the hub, one levels the straightedge at the trench, and one measures down with a tape measure and marks the pin in the trench.

Grade Rods for Final Grading
All the measurements considered so far were to the flow line of the pipe. The next thing that you must do is make up grade rods for

final grading or pipe laying. Let's assume the specifications call for a 0.30-foot crushed rock bedding under the pipe being laid. The pipe is 0.20 foot thick from the outer wall to the inner wall. Once the bedding and pipe thickness are known, a grade rod for trenching can be made. This could be 1 x 2 or a 1½-inch round wood or aluminum rod. Be sure the rod is long enough. Measure from the end of the rod up 8.50 feet and make a mark. The extra 0.50 foot is for the undercut. The thickness of the pipe plus the thickness of the bedding material is 0.50 (0.20 plus 0.30).

The same procedure is followed when making up a grade rod for grading the bedding material except that the thickness of the pipe would be added to the 8 foot measurement, making it 8.20 feet. The grade rod for checking the pipe flow line would be an even 8 feet because the 8-foot string line was set to the flow line. The flow line is the grade given by the surveyors on the grade stakes.

You have made three rods three different lengths for three different operations: (1) The rod for checking the grade behind the trencher or hoe measures 8.50 feet. (2) The rod for checking grade for the bedding material the pipe will be laid on measures 8.20 feet. (3) The rod for checking the flow line of the pipe after it is laid measures 8.00 feet.

When a T-bar or a bottom line is used, a straight rod is all that is needed. When checking grade with a side line, an arm must be nailed or screwed to the grade rod to reach the side string.

Above-Ground String Lines

Setting a string line for any above-ground operation is done the same way with one exception. When setting a string line for a paving machine, self-grading curb machine, or checking any above-ground trimming with a hoe, there will be fills on some of the stakes. If a 3-foot string line is being set, just add the fill grade given, regardless of what the fill is, to 3 feet. Remember that even though the grade is being set for a surface grade, you should keep four grade pins set up at all times so any mistakes are easily seen.

When setting a string line for curb boards or anything where the string is to be placed at a finished grade, the measurements given by the surveyors are used as is. If the surveyor's stake has a 1-foot

**Using plastic pipe instead of
string for smooth turn
Figure 4-12**

cut to the top of the curb, measure out the offset given and down
from the level straightedge 1 foot and mark the pin. If the ground
in front of the hub to the curb is higher than the hub, use an extension on the straightedge. The amount of extension added must be
computed with the cuts and fills given on the surveyor's stakes.

When setting a string line for forms of any kind, set up the pins
and string so the pins will be on the outside of the boards. Corners
are difficult to make when setting up a string line for a self-grading
curb machine. Many pins must be set to keep the string flowing
evenly around the turn. Otherwise the line sensor will cause the
machine to make a very irregular curve. You can solve this problem by using a 1/2-inch plastic pipe in place of string on the corners. This will make the corner much easier to set, will use less
pins, and will help the machine turn out a smooth corner. See
Figure 4-12. Here the operator is helping the sensor past the grade

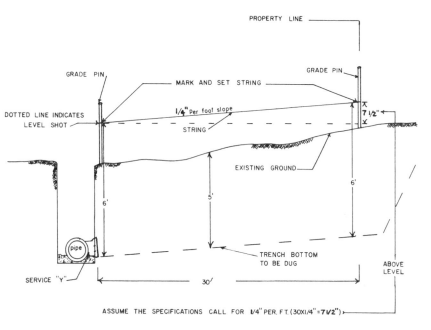

Sewer service grade line
Figure 4-13

pin arm. Notice the two sensors; one horizontal for elevation and one vertical for line.

Sewer Line Projects

On a sewer line project, when the locations of the sewer services are staked, the surveyor usually will not give a cut grade. When the sewer service is dug, the trencher or hoe operator may want a grade line to follow. If so, you can do it without the surveyor's assistance. Offset a pin from the service "Y" or tee on the sewer main and do the same at the service location stake set at the property line by the surveyors. From the bottom of the pipe, measure up and over 6 feet to the grade pin.

Look at Figure 4-13. Assume the trench is 5 feet deep and we have measured up 6 feet and over to the pin. Compute the amount of minimum fall required in the specifications and raise the string at the property line that amount. In this case, that's 7½ inches because the specifications call for 1/4 inch per foot slope. If you multiply 1/4 inch times 30 feet, you get 7½ inches. So shoot level from the front pin at the "Y" or tee to the back pin at the property line. Then raise the back string above the level the amount of fall desired (7½ inches) and mark. If the ground rises sharply from the "Y" or tee to the property line, tie the string above the ground the same distance at both the back pin and the front pin. The service line will then have approximately the same fall as the ground slope and will have more slope than the minimum 1/4 inch per foot.

When setting any grade line, be sure to check every figure. Avoid any distraction while figuring or marking grades. If you aren't quick with numbers, make it a practice to use a pencil and paper to do calculations or use a calculator. It's a common mistake to figure grades from a string of cuts and then come to a fill and figure it as a cut. Look at each grade stake separately and read it for what it is.

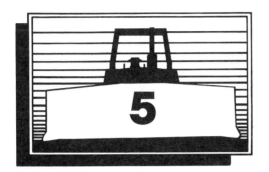

Laser Levels

Laser levels have replaced hand-held sighting levels on most large excavation projects. Laser levels are far more accurate over much longer distances. That helps the grade setter get more work done in less time. For example, from a single set-up, a laser level can be used to set grade for several lot pads.

The accuracy and range of a laser level makes grade checking much quicker. The grade setter can work well ahead of the equipment, spray painting marks for cuts and fills on the ground. While the equipment is doing that work, the grade setter can check what the grader has just trimmed. All this is done from a single set-up of a laser level.

Used correctly, a laser level helps the grade setter anticipate problems before the grader does the work. For example, with more grades finished faster the foreman can decide early if it's necessary to raise or lower the grade to balance the cuts and fills or to have a scraper haul in fill or remove excess. That saves grader time and reduces duplicated effort.

Laser receiver on rod with movable tape
Figure 5-1

Setting a Laser Level

Figure 5-1 shows the receiver end of a laser leveling set. Figure 5-2 shows the laser transmitter. The transmitter must be leveled precisely so it can project a level laser beam as it's turned through 360 degrees. When this beam is intercepted by the receiver, an arrow in the display window shows whether the receiver should be raised or lowered to intercept the laser beam exactly on center. When on level, the index line on the rod shows how far the beam is above grade.

To set a laser level like the one in Figure 5-2, rotate the adjustment wheels at the base of the level until the instrument is level. Some laser levels have bar-type levels on two sides. In this case,

Small laser unit sends out a level beam
Figure 5-2

both sides must show level. You can see the two bar-type levels in Figure 5-2.

If the laser unit you're using doesn't have bar levels, it may have a round circle-type level at the top or on one side. This type of level indicates level when the bubble is in the circle.

Most laser levels are self-leveling. If they're set close to level, they will precisely level themselves. They'll also automatically shut off when bumped off level.

Once the laser unit is leveled and turned on, you'll want to determine the height of the level beam being projected by the laser unit. To do this, take the rod, with receiver attached, to the nearest bench mark or known grade point. Set the base of the rod on the bench mark. Turn the receiver so it faces the transmitter. Slide the receiver up or down the rod until it intercepts the laser beam. It's on level when the receiver gives off a steady tone and only a bar appears in the window. Figure 5-3 shows a receiver signaling that

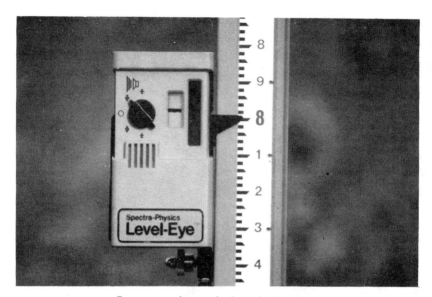

Laser receiver window indicating
it's level with laser beam
Figure 5-3

it's level with the laser beam. You can only see a bar, not an arrow pointing up or down. Set the elevation of the bench mark to the beam height on the rod. That's the elevation of the beam and can be used on all grades shot from that set-up.

Setting a Movable Tape

Use only a direct reading rod with a movable tape like the one shown in Figure 5-1. A tape like this can save a lot of adding and subtracting. It also makes errors less likely. Here's how to set the tape on one of these rods. Notice the nut and clip on the top right in Figure 5-1. The nut holds a plate with a hook used to attach the survey tape. Loosen this nut. Pull the hook-type plate outward so the tape can be cranked up and down. Crank the tape until the last three numbers of the elevation at the bench mark are at the pointer on the receiver. Then tighten the nut again so the tape can't move.

Once the tape has been set this way, one person can check any elevation within range of the laser beam. Just move the rod to the grade being checked. Slide the receiver up or down until the tone is steady and only a bar is visible in the receiver window. Read the tape at the pointer to find the elevation at that location. No adding or subtracting is needed!

Lets look at an example. Assume that 128.00 is the bench mark elevation. Figure 5-3 shows the rod setting at this bench mark. The movable tape has been cranked so 8.00 is at the pointer on the receiver. After resetting the hook-type plate so the tape can't move, we'll check the setting for accuracy. Facing the receiver toward the laser level, you should hear a steady signal. If the signal is not steady, make adjustments as needed.

The reading at the receiver is 8.00 in Figure 5-3. That was at the bench mark. Now we move the rod to another location, leaving the laser unit undisturbed at the same setup point. At the new location the receiver bar centers with the pointer at 2.58. See Figure 5-1. That's only the last three numbers of the grade, of course. The full elevation is 122.58. See how easy that is! No adding or subtracting is needed.

Look again at Figure 5-3. Notice that each black bar on the rod indicates 2 hundredths of a foot (0.02 feet). Bars (and the spaces between bars) are 1 hundredth of a foot wide.

If the beam being projected from the laser unit falls above or below the rod when it's on the grade you want to check, you'll have to move the laser unit to a higher or lower elevation. Before moving the laser unit, set a hub nearby and check the elevation of the hub. Then when the laser unit is moved, that hub can be used as a bench mark to reset the laser unit.

Laser levels are a big improvement over sighting levels. But there are some cautions. First, be aware that fog, dust and rain can deflect the laser beam, especially over longer distances. Reduce distances when there's moisture or dust in the air.

Note also that equipment working around the laser unit can disturb the laser beam. A vibratory roller working near the transmitter will make it hard to get a steady signal at the receiver. And be careful while checking grade on any job where more than one laser is being used. It's easy to make a mistake if a second laser level is projecting a beam close to the same level as your laser.

Larger laser unit sends level or sloped beams
Figure 5-4

Your receiver can't tell the difference between your transmitter and any other transmitter.

Types of Laser Levels

There are several types and brands of laser levels on the market. The small laser level in Figure 5-2 sends out only a level beam. It has an on-off switch and three bottom wheels for leveling. Notice the two level tubes at the bottom of the windows. This unit is run by four D batteries.

The larger laser level in Figure 5-4 can project either a level, declining or inclined beam. Just dial in the percentage of slope desired. The knob on the left changes the slope percentage. It's powered by a 12-volt car battery, which is set on the ground under the laser. The arrow on the laser's top must be pointing in the direction of the slope being transmitted.

A laser level that can project an inclined or declining beam is perfect for laying pipelines. Each length of pipe can be set precisely on the slope desired.

Notice that the laser transmitter in Figure 5-4 has window flaps that can be closed. In Figure 5-4 only the front window is open. There are three more windows, on the two sides and on the rear. When all four windows are open, the laser beam is projected through 360 degrees so leveling can be done in nearly any direction from the instrument at the same time. This is called a *twirl-type* laser level. A transmitter that projects a beam in only one direction is sometimes called a *stationary* level.

The twirl-type laser level has four small areas in 360 degrees where no beam will be received. The beam doesn't pass through the four posts that separate the windows. Try to set up a twirl-type laser so that a post isn't centered in the direction of your work.

Setting the Number on the Survey Rod

Remember, no matter how many numbers the bench mark may have, only the last three numbers will be used in the survey rod reading. That can cause some confusion if you're careless. For example, a bench mark in Los Angeles might show an elevation above mean sea level of 18.25. A bench mark in Denver may read 5218.25. There's a 5200 foot difference between the two bench marks. But the tape at the pointer of your rod would be 8.25 for both setups.

When checking elevations, be alert to situations when the fourth number of the elevation has been affected. Here's an example. If the rod is set at 8.25 and you check an elevation 1.70 higher, the rod will read 9.95. Using the Los Angeles bench elevation, that would be 19.95. The fourth number (reading from right to left) hasn't changed. It's still one.

On the other hand, if you're checking an elevation 9.95 lower than the 8.25 rod setting, you would read 8.30 on the rod. This time the fourth number has changed and the actual elevation is now 8.30, not 18.30. In the first example, you were adding 1.70 to the bench mark of 18.25. Now you must subtract 9.95 from the 18.25 bench elevation.

Be alert for elevations that are more than 10 feet above or below the reference benchmark. In that case, the fourth number of the bench mark has to change, even if the last three numbers stay the same. If there are large grade changes in the area being worked, avoid confusion. Do your calculations using the last four numbers of each elevation.

Using the Laser Level for Parking Lots

Laser levels are convenient when setting grade for parking lots. With a laser level it's quick and easy to reset grade hubs that have been run over by the equipment or need to be offset and reset after that area has been excavated.

Do some thinking before deciding where to set up your laser. There are two main considerations: First, you want a safe place where it won't be run over by equipment. Second, the setup point should offer good coverage of the job site. If the only place available for a setup is in a high-traffic area, place laths and flagging around the laser so it's less likely to be run down by accident.

If you have a large parking area that drains to swales or curbs, grading must be done precisely. Any low spots will collect water on the finished asphalt the first time the surface gets wet.

Start by doing the rough grading, if needed. Where deep cuts are required, set a 2 or 4 foot lath with the cut or fill marked on it. This is faster and more practical than driving hubs that will be excavated or filled over. Don't bother using the laser in areas where sweding the subgrade will be faster.

When rough grading is finished, establish a grid pattern on your plan of the parking lot. Grid lines should intersect about every 25 feet. Then lay out the same grid lines on the site. Have the grade setter drive a hub at each intersection point of the grid. Don't bother setting the hubs at the right elevation yet. That comes later. Next, figure out what the elevation should be at each hub so the entire area drains evenly from the highest point to the lowest. Mark those elevations on your plan.

When the entire grid is established both on the ground and on your plan, begin driving hubs to grade. The grade setter uses the laser to check the elevation at each hub. Drive each hub until the head is on grade when the pointer of the receiver is set at the right elevation on the survey rod.

Once the grade setter has set each hub to the desired grade and has placed a guard lath beside each one, the grade is ready to be fine trimmed.

If the parking area has several ridges and swales for drainage, plus curbs and planters around the perimeter, it may be faster to swede the grades needed from the swales and ridges rather than laying out a grid. Or, a combination of setting some grades and sweding some grades might work best. The grid only works well on very large surfaces.

Using the Laser Level for Apartment or Industrial Pads

On apartment or industrial pads, the surveyors will usually set four corners and some perimeter stakes. You'll have to set all the remaining grades needed for excavation.

Let's assume the pad is 300 feet long and 150 feet wide. The scraper operators can't be expected to hold a level grade over that distance. Grade control stakes will be needed across the pad when the cut or fill is within 6 inches of the finished grade.

With a laser level, the grade setter can check grade at any point on the pad from a single setup. When excavation is getting close to finished grade, there are three ways to inform the operators of how much cut or fill is needed. First, the grade setter can relay the information directly to the operators by hand signals. Second, he can paint cuts or fills on the ground. Third, he can set crows feet across the pad with the cuts or fills indicated on them.

If the grade setter has time to stay with the scrapers, he should look for areas that are on grade and paint an on-grade symbol on that spot. Remember, the on-grade symbol is a circle with a line through it. Setting lath is usually the best method when a fill is being made. The scrapers dump faster than they cut. They may not have time to read symbols painted on the ground. If the grade set-

**Using a laser to check pad grade
ahead of the grader
Figure 5-5**

ter has other areas to check and can't stay with the scrapers, use
laths to indicate the grades.

Once the pad is nearly on grade, it's ready to be trimmed with a
grader. The grade setter then works ahead of the grader, painting
trim or fill symbols on the ground. After the grader has made a
pass, check to be sure the trim or fill was made correctly. Figure
5-5 shows a grade setter checking ahead of the grader. Note that
several grade setters can work at different elevations off the same
laser level if the beam isn't too high or too low for the receivers.

With a laser rod, you don't use a boot or offset as you have to
when checking grade with an eye level. Instead, the grade setter
marks with a pencil the elevations he needs on the back of the information lath. For example, if 2 inches of asphalt and 6 inches of
aggregate are required for a parking section, the three elevations

marked on the lath might be 135.00 (for finished asphalt), 134.83 (for finished aggregate), and 134.33 (for subgrade elevation). During excavation, the subgrade elevation is all that's really needed on the lath.

Using a Laser on a Road Job

On a road job, the elevations of ditch, shoulder, and centerline subgrade would be marked on the lath. Each time the grade is checked at a station, the grade setter can see on the lath the elevation he should have on the survey rod. It helps if the finished elevations for each point are shown on the plans. That makes it easy for the grade setter to subtract the section he wants to cut from the finished grade shown.

If the surveyor sets a hub and information stake with a cut of 0.70 and you want to check subgrade with the laser rod, here's how to do it. First, set your survey rod on the hub in question. Then subtract the cut and the road section depth. Assume the survey rod reading on the hub is 120.50. The surveyors have a 0.70 cut to finished A.C., and there is a 0.75 section. Subtracting cut and section from 120.50 gives the grade setter an elevation of 119.05 for subgrade to read at that point. He'll mark 119.05 S.G. on the lath so he doesn't have to refigure it every time the station is checked.

If the surveyors have a fill rather than a cut to the finished grade, the fill must be added to the rod reading before the section is subtracted. For example, if we use the same rod reading, 120.50 elevation with a fill of 0.70 and a 0.75 section, the rod reading for subgrade should be 120.45. That's 120.50 plus 0.70 minus 0.75, or 120.45. That subgrade elevation should be marked on the lath to save calculations later. Remember, a road or parking lot section is the thickness of A.C. and aggregate between finished grade and subgrade.

When excavation of the subgrade (level of the compacted native earth) is complete, it's time to begin placing aggregate. The grade setter has to figure the correct elevation for aggregate at each station. He'll add the thickness of the aggregate to be placed to all the subgrade elevations he should have marked on the lath while excavating to subgrade.

A grade setter experienced in checking grade with a laser may need surveyor's stakes only to establish distances and line. He may prefer to compute all elevations from the plans.

Set the rod elevation from an accurate grade reference such as a bench mark. If a bench mark is not available, use a point on the plans with an elevation indicated, such as existing drain inlets, curbs, concrete slabs or walls. If they haven't been disturbed, survey hubs for pads, curbs, or pipe are good reference points. Once the elevation is set, check the setting at a second and third known elevation point to be sure the setting is accurate.

If you're using a laser level for grade control on a subdivision, road or parking lot job, be sure to ask the surveyors to set an accurate bench mark somewhere on the site.

Setting a Laser for Trench Work

When setting up the laser level for trenching, select a location where it won't be bumped by the equipment. In Figure 5-6, the laser is set several feet from the work area, facing down the trench. The hoe and grade checker can both receive the beam from a twirl-type laser. If you're using a laser that transmits a stationary beam, set it at the centerline of the trench. For a laser with a twirling beam, it needn't be on the centerline of the trench because it's only used for grade and not line.

The advantage of the twirling beam is that it can be used while laying or trenching on a radius. Also, being offset from the centerline of the trench, it can be used for trenching and laying simultaneously. The pipe laying operation can go on while the laser guides the work of the trencher or hoe.

Setting the Percentage of Slope on the Laser

There are two ways to set the percentage of slope on both the fixed beam and twirling-type laser. Some transmitters have a calculator-type key pad. Just punch in the slope you want with a plus or minus grade. On others you dial the slope with a knob. Trenching

Place the laser in a safe location
Figure 5-6

or pipe laying is kept on line with a string or chalk line when using twirl-type lasers.

Setting Up the Laser for Trenching

To set up a laser for trenching or laying pipe, set the rod on a hub set by the surveyors. Slide the laser receiver up or down the ruler until the receiver gives an on-grade signal. Using a ruler, measure the distance from the hub to the pointer on the receiver. That's the height of the beam above the hub. Now add that measurement to the cut the surveyors have marked at that station. Measure that distance on a 1 x 2-inch rod and draw a line. Now clamp the laser receiver to the 1 x 2 with the pointer level with the line just drawn. The rod is now ready to check the grade at that station for the flow line of the pipe.

Let's try an example. Assume the measurement from the hub to the receiver point is 3 feet and the cut on the survey lath is 7 feet. Adding the two measurements you get 10 feet. Now measure up the rod 10 feet and clamp the laser receiver at that point. The rod is now set. Drop it in the trench and see if you get an on-grade signal.

Notice that using this method the pipe invert (flow line) elevation isn't needed. You only need to know the distance from hub to the receiver and the cut indicated on the survey lath. Be sure to follow this procedure at three stations to be sure all three points give a 10-foot reading. If they don't, there's a mistake in the slope dialed into the laser.

Remember that an undercut is needed for bedding material and the thickness of pipe. You have to add this undercut to the rod elevation or grade pole length. And be careful to clamp the receiver tight enough so it won't slip. Get in the habit of placing a mark on the grade rod when it's set so any slipping will be noticed.

The laser level simply supplies a level beam to set grade from. It's the same as checking grade from a string line except that setup goes much faster.

Laser Levels on Equipment

Laser receivers are often mounted right on the equipment on larger projects. Consider this if you're grading farm land or large storage pads or even long levees or channels with a constant slope. Laser control is also useful for subdivision excavation. Lot pads and streets can be excavated by dialing the elevation required in each area.

A laser-controlled compactor can save a lot of time. A hydraulically-controlled laser unit can operate the equipment controls automatically. In Figure 5-7, the scraper is equipped with an on-board laser system. You can see the laser receiver extending above the paddles on the scraper bowl. On the bottom left of the front window, there's a laser control box where the operator sets his elevation and gets his readings. On this particular unit, the operator dials in the elevation he wishes to cut, then manually operates the controls. The laser unit will read *Hi* or *Lo* until the

A scraper with on-board laser equipment
Figure 5-7

grade dialed in is reached. Then the unit will read *On*. When it reads *On*, the operator knows he's on grade.

When using equipment-mounted receivers, the transmitter will have to be placed on a platform so it's high enough to reach the target on all equipment in use. Figure 5-8 shows a platform with steps so the grade setter can adjust the equipment easily. No matter how many pieces of equipment or what size, the target arm on each must be long enough to intercept the laser beam. In Figure 5-8, the platform's tires are off the ground and the mast the laser unit sits on is well braced. This keeps the unit steady if the wind blows, and assures a steady signal.

Using a Laser Mounted on a Scraper

Here's how to get set up when lasers will be mounted on a scraper. First, mount the transmitter on a platform in an area where it can

Laser mounted on platform
Figure 5-8

cover all the job. Next, set the cutting edge of the scraper on a bench mark or surface of known elevation. Raise or lower the receiver pole until the laser unit aboard the equipment gives an on-grade signal. Then dial the elevation of the bench mark into the on-board laser. The on-board laser is now calibrated. From this point, procedures vary with the model or brand of laser unit used.

In general, the elevation of each cut or fill must be dialed into the on-board laser. As the operator gets to the cut area, he sets the elevation that area must be cut to. When he approaches a fill area, he sets the elevation to be filled to. The on-board laser unit will operate the bowl of the scraper up or down so the cutting edge will be at that elevation. If it isn't an automatically controlled unit, the operator must manually operate the scraper bowl. Every system has an override so the operator can take over, mainly if the cut or fill is too large for using the system.

Laser receiver mounted on hoe arm
Figure 5-9

When on-board lasers are used, the operators must have a small plan of the lots and streets, and their elevations, so they will know what elevation to dial into the laser unit. The laser may be mounted on a grader for pad trimming also, and used in the same manner.

Using a Laser Mounted on a Hoe

Set up the laser on a hoe or trencher the same way. In Figure 5-9, the laser receiver is attached to the hoe arm. An indicator panel in the operator's cab shows when the grade is reached. No string line is needed for checking grade.

When the trench bottom is on grade, the hoe operator sets his bucket on the trench bottom. It's important, of course, that the angle of the digging arm and hoe bucket be the same every time the grade is checked. Then the operator slides the receiver up or down the tube until a level reading is indicated on the panel in the cab.

Whenever the laser beam intercepts the receiver, it sends a reading to the operator indicating cut, fill, or on-grade. That makes checking grade a continuous process, even as the hoe continues to dig.

Laser levels are like any other sophisticated equipment. If used properly for the right application, they speed up production. Used incorrectly, the results will be unpredictable.

Other On-Board Control Systems

There are many grade control systems that aren't laser operated. Many graders are equipped with a system for slope control that doesn't need a laser beam, wire, or string. To set these units for work, the cutting edge of the blade is first set level. The control box mounted in front of the operator is set at zero percent and locked in. Once this is done, the operator can dial in any percentage of slope he wishes to cut. The unit will automatically control the cutting edge on the right, left or both sides to the slope dialed in. If too much material is being cut, the operator can override the system and take over manually.

This system works well on road jobs where a crown or super elevation is being trimmed. The operator can manually control the cutting edge over the guinea along the road edge set by the grade setter. By dialing the percentage of slope needed for the opposite side cutting edge, the slope control automatically controls that side while the operator controls the guinea side. See Figure 5-10.

Once the operator has finished his trim over the guineas, the slope control has cut the opposite side to the exact slope dialed in. The operator now moves over and manually holds the cutting edge to match the outside pass that was just cut. With the slope still dialed in, the cutting edge on the opposite side is again automatically controlled to cut the correct slope. A wide road section can be trimmed in this manner without intermittent stakes.

The on-board slope control saves a lot of the time that normally would be spent setting hubs and retrimming. It can save hours or days on large building pads that are level or have slight slopes. One side can be set to grade and then the rest of the pad is graded by dialing a slope or level on the control system used.

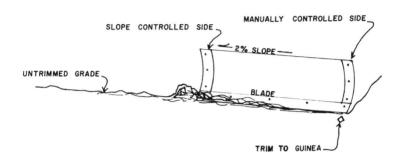

SLOPE CONTROLLED SIDE

MANUALLY CONTROLLED SIDE

2% SLOPE

UNTRIMMED GRADE

BLADE

TRIM TO GUINEA

A   First pass

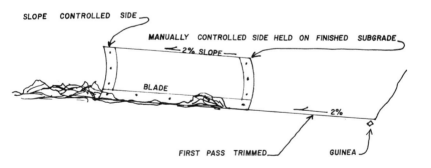

SLOPE CONTROLLED SIDE

MANUALLY CONTROLLED SIDE HELD ON FINISHED SUBGRADE

2% SLOPE

BLADE

2%

FIRST PASS TRIMMED

GUINEA

B   Second pass

**Using a slope control unit
Figure 5-10**

The most important point in using slope control is to cut the first pass accurately. If the first pass has any **humps or** dips between guineas that were set or is left high at the guineas, each succeeding pass will come out the same way.

There are a variety of new grade control features for graders. Besides the laser system that's controlled with the laser beam, there's a sonar system that will take grade from any surface such as string line, curb or existing grade. These systems help to improve the quality and speed of the work being done, but they're expensive. Examine every system and choose the one best suited for your needs.

Shop Talk

Since the first edition of this book was published in 1978, many types of highly productive heavy equipment have been introduced. This chapter is intended to expose you to slip-form pavers and curb machines, profilers and reclaimers that could help cut your costs on future jobs.

Curb Machines

Curb machines for various shaped curbs and barriers are now used regularly. They're so efficient that they've replaced conventional form setting on all large jobs. Figure 6-1 shows a slip form concrete machine easily pouring a 42-inch vertical barrier. Figure 6-2 shows a concrete machine running on a pretrimmed grade. Notice the consistently light windrow being left, and the very smooth concrete pour.

In Figure 6-3 you see a crew pouring an extruded curb at the edge of the asphalt in a parking lot. Epoxy has been spread on the

Slip form concrete machine pouring a large barrier
Figure 6-1

**Concrete machine running on a pretrimmed grade
Figure 6-2**

**Curb machine with auger
Figure 6-3**

Self-propelled curb machine
Figure 6-4

asphalt, ahead of the curb machine, so the curb won't slide when bumped later. This curb machine has an auger at the bottom that propels it forward while it pushes concrete into the small form at the back. The wheels are raised just enough for the machine to drag slightly on the pavement. One man steers in front while two men behind pour the concrete.

Figure 6-4 shows a self-propelled curb machine pouring a 14-inch barrier curb in a parking lot before the aggregate base is trimmed or the paving is done. Note the grade control sensors for line and elevation.

Slip Form Pavers

All concrete roadways are now poured with concrete slip form pavers. In Figure 6-5 a slip form paver follows a spreader. It leaves

Slip form paver following a spreader
Figure 6-5

Tube-type finisher gives a troweled finish
Figure 6-6

a smooth uniform mat of concrete while moving forward, following the string line for grade and line. When a spreader does the initial spreading ahead of the form paver, it increases the paver's production rate. The slip form paver can pave more than one mile of 24-foot wide road in one shift and still get excellent profilograph readings. A profilograph is a straightedge on wheels. The inspector rolls it along the road surface to detect any areas that may be out of tolerance.

Float machines and a tube-type finisher following behind the paver can eliminate virtually all hand finishing on a concrete highway job. See Figure 6-6. The tube finisher is followed by a machine that tynes (scratches small grooves into) the surface and sprays curing compound on it.

Most slip form pavers have a spreading auger that distributes the concrete the width of the paver, a primary strike-off bar, a row of vibrators, a secondary vibrating strike-off bar and one or more screeds for the initial finish. Then a large float device floats and edges the mat of concrete before it leaves the machine. The machine can be adjusted so it leaves a crown in the concrete mat if needed. The operators and mechanics that work with this kind of equipment are usually trained by factory-trained personnel who stay on the job site until the contractor's personnel can take over.

Profilers

Profile machines are now available for resurfacing freeways, runways, secondary roads and parking lots. These machines are extremely cost efficient in remote areas. A profile machine can mill off 4000 tons of road surface a day to a tolerance of 1/8 inch. Grade and slope are usually controlled by a wheel-type ski sensor. The old asphalt that's milled off can be loaded into trucks by the conveyor on the profiler and hauled to the asphalt plant for recycling. Recycling is usually done by mixing a percentage of milled asphalt with virgin material. Concrete can also be milled.

Profiler milling old asphalt
Figure 6-7

Figure 6-7 shows a profiler making a second pass and milling 2 inches of old asphalt. The profiler loads the asphalt on the truck. Notice the small sensing wheels on each side. They send readings back to the controls that automatically control the elevation being trimmed.

Rows of carbide teeth on a drum, spinning at a high rate of speed, do the cutting. A spray system sprays water to cool the teeth and keep the dust down. In Figure 6-8 you see a profiler with a second cutter-type grizzly for sizing. A grizzly is a screen or bars that let only the required size of aggregate through. This profiler, hooked to an oil truck, mills 4 inches of old surface. Spraying liquified asphalt oil into the milled material produces a cold mix asphalt that is ready to be spread.

Reclaiming Machines

The reclaimer works like a large rototiller connected to an oil truck. As it rototills the asphalt or cold mix and aggregate together, a controlled amount of asphalt oil is added. See Figure 6-9. Usually more than one pass is needed to pulverize the material well. The road can then be graded, rolled, and a surface course of asphalt concrete or cold mix paved over the oiled base.

On a rural road where traffic is light, no surface course may be needed. The oiled base may be simply rolled and chip sealed. This reclaiming method saves the time that it would take to rip and pulverize with a grader and compactor. Making several passes with a compactor won't pulverize the asphalt surface as well as a reclaimer. Note that reclaimer machines don't have the grade and slope control capabilities of a profiler.

Many miles of primary and secondary roads in the U.S. need resurfacing or rehabilitation. Profilers and reclaimers should be very busy for the next few decades.

Other New Equipment

Trenchers equipped with carbide teeth are now able to trench through concrete, asphalt and rock. This eliminates jackhammer

Profiler hooked to an oil truck
Figure 6-8

Reclaimer pulverizing old surface
and base and adding oil
Figure 6-9

work or blasting in many cases. There are also new and improved hoe attachments like hoe rams for breaking concrete or rock, or hoe pacs for compacting narrow or restricted work areas like trenches. Stay current on the new machinery that's available.

Plastic pipe makes pipe installation much faster because it's light and comes in longer lengths. A variety of joints such as slip ring, glued couplings, and straps make adding fittings easy. It's now used for drains, sewer, and water lines. Pipe cutters aren't needed. A cut-off saw is used for most pipe cutting now because it's faster.

Communication Is Important

There are no secrets in the excavating and paving business. The work is done outside. Anyone that wants to can watch. If you're a foreman or superintendent, keep an eye on the competition. If a competitor is using new techniques or more productive equipment that gives them an edge in the market, consider making the same change in your company.

It's my opinion that every foreman should know the estimated cost of each major part of the job, both for labor and equipment. This lets him compare estimated productivity with actual productivity. If actual daily costs are running more than estimated costs, something is wrong. The foreman may be able to make equipment or employee changes to lower the daily costs before a loss becomes inevitable.

Most companies are hesitant to give the foreman this information. I believe this is a mistake. Human nature being what it is, most crews will try to beat the estimated productivity rate if they know what it is.

I also feel a major cause of cost overruns is lack of communication between the estimator, superintendent, and foreman. If they would spend some time together walking the job and discussing the way the job was bid and potential problems, the job would begin more efficiently.

Here's an example. Consider an excavation with excess dirt to be exported to a dump site. The superintendent and foreman may get the plans and set up a crew using scrapers to stockpile the excess dirt. Then they plan to load it into trucks with a loader and haul it to a dump site. Maybe the estimator figured the job entirely differently. Maybe, because of the depth of the excavation and the restriced work area, the estimator decided that using a large hoe to excavate directly into trucks would be more cost efficient. He bid the job accordingly. He may have based his estimate on a haul to a nearby dump area rather than the one that's actually being used. If the foreman and superintendent aren't given this important information, all the research done by the estimator may be wasted.

Safety Is Important Too

Safety is an important part of every construction business. Never compromise safety for production. Injuries caused by negligence can lead to fines and even jail terms or both. A contractor who continues to have safety violations and accidents will eventually be overwhelmed by his liability insurance costs.

Protecting the public passing through or around the project is equally as important as protecting your crews. Be sure to take time during and after each work shift to check signs, barricades, and arrow boards. It may save thousands or millions of dollars in damage awards. Any time you work, work safely!

Excavating
Subdivisions

This chapter explains how to excavate and grade subdivision streets and lots. Most jobs should proceed in the sequence outlined in this chapter. As in most excavation work, planning is important. Never let the equipment stand idle while grades are being set or while making a decision on where to start the next cut or fill. Try to solve problems before they hamper production. Make sure a water permit is obtained, if needed, and all existing underground utilities are marked.

Study the plans and stakes carefully before starting. Take the time to analyze the type of equipment required. This will eliminate reduced production that results from poor equipment balance. If you're using a paddle wheel scraper, know the soil condition. Determine whether a paddle wheel with ripper teeth can cut the material or you need a dozer ripping to help. Check the soil moisture and distance to the water source. This determines the number of water trucks needed and the truck size. Remember, the size of the equipment should be determined by the size of the job.

Problems may arise that aren't covered in this chapter. Most of these are covered elsewhere in this book. Check the index for additional information.

Staking the Property

The surveyors lay out the subdivision property by running a row of stakes down the center of each street. On each stake, they'll note that the stake marks the centerline, write the station number, and give the cut or fill to the finished road grade. At the base of each information stake, there's a hub driven in the ground. The grades shown on the information stake will be taken from this hub. It may have a small tack somewhere on the top surface which is the reference point on the hub. The surveyor computes the grades at that station from that point.

The lots are usually staked at the property line. Front lot stakes will probably be driven at the setback distance. Back of lot stakes will be at the back corners of the lot. These stakes may show offset distances to building pad corners. If you're dealing with extremely large lots and big elevation changes between lots, you may need a different staking pattern. If the lots are large, they may have additional intermediate stakes. Each stake will have the information needed to cut the lot to grade at that point written on it.

In some cases, one stake and one hub can be used for two adjacent lots. Then one side of the stake will read *lot 22* and the amount to cut or fill. The back side of the same stake reads *lot 23* and the cut or fill. It's very common to have a fill for one lot and a cut for the adjoining lot. Usually, the surveyors tie different colored ribbons to the tops of the stakes so the construction crews can easily distinguish the lot stakes from the street stakes. Front and back lot pad stakes may have offset distances on them. Section A in Figure 7-1 shows front lot and street centerline staking.

The surveyors may stake back of sidewalk elevations, eliminating the centerline stake completely. If this is done, the grade setter must compute the centerline grade from that point by referring to the street section on the plans. Many contractors prefer this method of staking streets because it eliminates

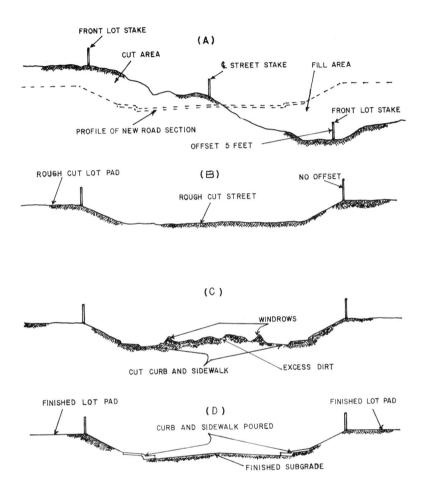

Four steps in street staking and grading
Figure 7-1

centerline stakes that must be offset or worked around. If you
have a preference, make it known before staking begins.

Starting Excavation

To start evcavation, first strip the area of all vegetation by either
disking it under or scraping it off with earth-moving equipment

Self-loading scraper removes strippings
Figure 7-2

and dumping it in a designated area. Shanks and teeth should be removed from the scraper's cutting edge before stripping to reduce the amount of soil taken with the vegetation.

In Figure 7-2, a self-loading scraper is making a final pass to remove the strippings. A water truck wets the next area to be stripped for dust control. The strippings are usually dumped at the back of the lots. Don't place any vegetation in the building pad area because most anything that grows is unsuitable for compaction.

A soils engineering firm usually directs the contractor to the areas available for dumping vegetation. The engineer also checks lot pads for expansive soils such as clay. If the engineer thinks a lot pad has too large an area of expansive soil, he may direct the contractor to remove the expansive material and compact suitable soil back into that area as an extra work item. The soils engineer usually tests all fills at various levels as they are being filled.

Setting Street Stakes

After the entire job site has been stripped of all vegetation, the grade setter can set his street stakes. The street width is indicated in the road section on the plans. The grade setter should run two rows of stakes on each side of the street at the back of the proposed curb or sidewalk, parallel with the centerline stakes and at the same stations. For example, if the plans indicate the road is 48 feet across from the back of sidewalk to the opposite back of the sidewalk, the grade setter measures 24 feet from the centerline and drives a stake at a right angle to the centerline. He does this at each station and on each side of the centerline. This defines the street area to be cut before the slope starts up or down to the lot pads.

The grade setter may also have to transfer the centerline grades set by the surveyors to his stakes at the back of the proposed curb or walk. If the excavation crew is experienced, the lath won't have to show the elevation. The grade at centerline is the only grade indication needed. The grade setter can check at each station and tell the scraper operators how much cut is left to make. The grades set by the surveyors indicate the height to the finished road surface. The grade setter therefore must add the thickness of the road section to all cut grades and subtract the same thickness from all fill grades. Then all the cuts and fills become subgrade cuts and fills rather than finished grades.

Underground Utilities

Remember to consider underground utilities while setting up subgrade cuts and fills for a subdivision street. Placing drainage and sewer lines and hauling material in for bedding or the initial backfill over the pipes will create a surplus of soil. If this excess isn't compensated for during the street excavation, it'll hinder the crew that returns to cut the curb grade after the utilities are in. The foreman must determine the amount of excess anticipated from the utilities and undercut the subgrade to allow for that excess. A good rule of thumb to follow is to undercut the centerline the amount required for the crown.

For example, if the road will be 24 feet wide from lip of curb to lip of curb with a 2 percent crown at the centerline, add 0.24 feet to the subgrade cuts at the centerline. Cut a flat grade and allow for the excess utility dirt to make the crown later. If the soil is very sandy, this practice of ignoring the crown won't apply because sandy soil shrinks more when compacted. If 10-inch or larger sewer or drain pipe will be laid in the street, more undercut will be needed. Remember the amount of undercut is determined by the size of pipe and the amount of import backfill material required for trench backfill.

Sections B, C and D in Figure 7-1 show a common method of cutting a street subgrade. But if you have a crew that has worked together on the same type of excavation several times, they may have a slightly different method which they feel works better for them.

Figure 7-1 assumes the subgrade for curb and street are the same. Often they're not. The street subgrade may be deeper. In this case, there should be a vertical cut at the front of the curb to the subgrade of the street that's deeper.

With the newer models of curb trimmers, some contractors are undercutting the sidewalk and curb grades and leaving the street subgrade slightly higher. This leaves an excess of dirt in the street after the underground utilities have been placed. When the crew returns to cut sidewalk and curb grade, they use the excess to fill the sidewalk and curb grade. In hard soils this eliminates ripping the sidewalk and curb grade, rolling it out before adding water, then rolling it back for compaction.

Every foreman has a slightly different way to cut a street subgrade. The main thing is to determine which way is fastest for you and will undercut enough so there is a slight excess when the grade is trimmed later. It's better to have a slight excess than to be short of dirt when the crew returns for the final subgrade trim after utilities and curbs are in.

The actual excavation can start once the grade setter has all the street stakes and grades in place with the proper figures added to cut the subgrade. Blue survey tape should be tied to all fill stakes to indicate the point the fill should reach, or the fill portion of the lath can be sprayed with red paint. This makes it easier for the equipment operators to see the fills needed. If the equipment is equipped with laser units, each operator should have a reduced scale of the plans so he knows what elevations to dial in while cutting or filling lot pads. The alternative is to staple a 3 x 5 card with the pad elevation on the lath for the operator.

Starting the Excavation

Here's how to start the excavation. Study the cut and fill stakes closely and figure out a cut and fill pattern to eliminate any long hauls. Unless there are extremely large cuts in the lot pad area, it's usually best to start at the front lot stakes and work toward the center of the street. Keep the cuts as even and level as possible. With each pass the scraper should move more toward the center of the street. The steepness of the front lot slope determines the amount the scraper will move toward the centerline on each succeeding pass. The excess dirt should be hauled to the nearest lot or street area needing fill.

If the front lot stakes show a fill and the street stake requires a cut, make the street cuts and continue filling along the front of the lots until the fill grades marked are reached. If the front lot stake has been offset toward the street, undercut the lot slope on the street side of the stake. The amount to be undercut depends on the offset distance and the steepness of the front slope. See Figure 7-3. Once the lot fills and cuts have been made, the front lot stakes should be set at the actual front lot setback line. Then a scraper can straddle the ridge of dirt left and blend the front lot slope with the lot and street.

Fill the front portion of the lots first. Then work across the lot from the front stakes to the back property stakes. Attempt to set up the earth-moving operation so the spread can work from one cut area to the next. The fill areas between cuts should be completed as the dirt operation moves ahead. This eliminates any doubling back once the equipment has left the area.

Good Operator Procedures

The difference between a good operator and a poor one is usual-

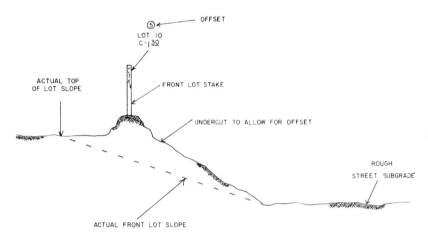

Front lot sloping
Figure 7-3

ly obvious when the excavation nears completion. Be alert to the
following points when excavating subdivisions:

1)   A good operator is always aware of grade changes in streets
and lot pads. When the area being excavated gets close to grade,
an alert operator will make smaller cuts, usually lowering the bowl
to cut 0.30 foot or less. While moving ahead, he raises or lowers
the bowl to adjust for lot steps or summit changes. Moving from
the area, he trims into a deeper cut, lowering the bowl promptly to
fill faster. This is also true if laser equipment is used. A good
operator knows when to use the laser control and when to operate
manually where the laser would be insufficient.

   An inexperienced operator or one who doesn't understand how
the finished job should look will usually skip areas where only a
slight trim is needed and will cut in only where a deep cut is required. He feels more secure working in the obvious areas. But it's
expensive to come back later for the small cuts that were missed
during the initial excavation.

   It's important that the grade setter let the operators know where
the summits and grades change. This may be done verbally, by

hand signals, or by placing colored ribbon on the lath at these points. Undercutting summits while excavating streets is a common error.

2)   The operator should remember that a road surface is never flat. It will always have a crown, side slope, or summits, or a combination of these. The same may be true of lots. There may be a difference of one foot or more from the front of the lot pad to the back property line, or from one lot pad to an adjoining lot pad.

3)   If possible, the foreman should start the lot excavation in the direction of the downward steps. This will help loading because it's easier to load working downhill. Also, there's less chance of undercutting lots because each succeeding lot will be lower. If there's a large drop between lots, cutting crossways may be necessary because the rams on the scraper may not extend down far enough to make the deep cut. Usually the top of a lot slope starts 1 to 2 feet past the property line and the slope extends onto the lower lot.

4)   Excavate the streets and lots together, especially if there are steep grades between the streets and lots. It's a mistake to concentrate on the streets before moving to the lots if there are deep cuts to make. If the street has a 6-foot cut and the scrapers excavate only the street, you'll have a 6-foot vertical wall on each side of the street. That makes extra work when lot front slopes are cut. It's not easy to work scrapers along a 6-foot vertical bank. Avoid this by excavating front lot slopes and streets simultaneously.

5)   It's equally important that a scraper operator be aware of elevations in the fill area when dumping. A thin, even dump is usually best. Thin lifts of soil mix quickly and evenly with water and are easier to compact. If the scraper is equipped with a laser unit, the operator should manually operate the controls while dumping or cutting until the pads are close to the elevation required.

Compactor operator mixes and compacts material
Figure 7-4

Sometimes it's hard to lay down thin lifts of fill. For example, in a tight corner the scrapers can't maneuver. In this case, the scraper operator should dump in a pile short of the corner. Then the water truck adds water to the dry material as the compactor operator mixes it and dozes it into place. That's what's happening in Figure 7-4.

If the compactor has a laser unit, the operator should dial in an elevation 0.10 foot above the elevation required. This gives the grader a slight trim for pad trimming.

6)   An experienced compactor operator can tell how much water is needed and where he needs more fill. The compactor operator can use hand signals to the water truck driver and scraper operators to direct placement of water and fill. It's very important to have a compactor operator who is knowledgeable about compaction and the amount of water needed.

While rough grading the lots, it's best to leave them 0.10 foot above grade. There'll be some shrinkage during the lot trimming and rolling, and trimming is much faster if the grader operator has a slight excess to work with.

Watch for Surveying Errors

If, at any time during excavating, the stakes seem to be off grade or it looks as though dirt from the cuts isn't going to balance with the fill needed, call the engineering firm that staked the project. They'll be happy to recheck any stakes that seem to be off. Even experienced survey crews can make a mistake on a station or two. Also it's common for surveyors to adjust a few lot grades to get the dirt to balance. The important thing is to watch for either of these two problems and correct them early. The front lot slopes should be left concave to accept the excess lot trim dirt and foundation dirt from the building.

Fine Grading

After all the rough cutting has been completed, the lots should be fine graded to the tolerance called for in the specifications. If the specifications call for a lot tolerance of 0.20 foot plus or minus, you can expedite the lot grading by averaging the cuts and fills. For example, suppose every second lot rises 0.10 foot. The eighth lot would then be 0.40 foot higher than the first. Raising the grade on the first lot 0.20 foot and lowering the grade on the eighth lot 0.20 foot will create a level grade through the eight lots but still stay within the 0.20 foot tolerance specified. This speeds up lot grading because the grader operator has a longer run without a grade change. Figure 7-5 illustrates this situation.

If you take care during the excavation and rough grading of the lots, you can complete the fine grading by using a grader and steel drum vibratory or rubber tire roller. The small amount of excess dirt generated can be "lost" at the front of the lot between the building pad and street. If lot pads were over-filled slightly, you usually won't need a roller when fine trimming pads.

**Grader can boost production
by averaging the pads
Figure 7-5**

Except for side, back, and front slopes, the entire lot pad between property lines may be graded level. The swales are usually cut by the landscaper when the building is nearing completion.

When the lot grading has been completed, place the underground utilities. Then the surveyors return to stake the curbs and walks. If there are no curbs, set ditch and street grades. All the grades given are finished grades. If there is a ditch rather than curb, cut the ditch and road grade. If curbs and sidewalks are required, a grader and grade setter cut the curb and sidewalk grades. The grade setter must add the thickness of the concrete curb or walk to the cuts and subtract it from the fills. He must also remember to take into account any rock material that might be called for under the curb.

Refer again to Figure 7-1, section C. If a trimmer is used ahead of the concrete machine, then the curb and sidewalk subgrade

Pretrimming curb and sidewalk grades
Figure 7-6

must be left from 0.10 to 0.30 high so the trimmer will have excess to trim. In this drawing you can see a windrow of dirt on each side of the street. A windrow is only left if the curb and walk are to be formed and hand graded. The windrows would be used to back up forms and for grading material. When a concrete curb machine is used, the street excess should be flattened.

In Figure 7-6, the curb and sidewalk grades are being pretrimmed ahead of the concrete machine on a large job to boost production. Because the grade was left high, it's being trimmed with ease. The conveyor deposits the excess in the center of the street.

Once curb and walk grades have been cut and the curbs and walks poured, the street compaction and fine grading can proceed.

Setting the Final Street Grades

The grade setter takes his final street grades from the curbs by computing the percentage of slope to the centerline. For example, if the street is 30 feet wide and has a 2 percent crown, the center would be 0.30 higher than each curb lip at the finished grade. This is computed by multiplying half the street width to centerline by 2 percent.

There are three ways the grade setter can set his grade. First, he can shoot it with his eye level or laser. Second, he can stretch a string from lip of curb to lip of curb and measure down. Or third, he can use three laths and sight across the tops. This last method is called *sweding*. The center lath (or swede) is shortened or lengthened to give the desired crown. All three methods are accurate. The string line and swedes are faster but they take two or three men rather than one. In some cases, the specifications call for the surveyors to restake the grade each time.

By this time a small excess of dirt is built up from the curb grade cutting and excavation for utilities. This dirt can be dumped on any low areas at the front or back of any lot. House pads have already been fine graded and these grades can't be changed. When the excess dirt has been removed, the road should be approximately 0.10 to 0.05 foot above grade, depending on the soil type, to allow for shrinkage during compaction.

For good compaction, the road should be ripped up 6 inches deep, watered, mixed thoroughly and then compacted. Usually a compaction test of 95 percent is required for the top 6 inches. After the compaction process has been completed and the road reshaped by the grader, centerline hubs can be reset. Take the grades from curbs by using swedes, reshooting, or string lining. Now the street can be fine graded and rolled.

The subgrade is now ready to be paved or rocked, whichever is required in the specifications. See Figure 7-1 D. If it'll be paved, no further grade setting is required. If rock will be placed, however, the grade setter sets a row of short laths about 2 feet high every 50 feet down the centerline, marking each stake for the finish rock grade. This gives the grader operator something to work to while he is spreading the rock.

After the rock is in and compacted, the grade setter sets the hubs or the surveyors reshoot the grade at the centerline for the fine grading. When the grading has been completed and compaction tests passed, the rock is oiled if necessary and the road paved without additional grade setting.

Finishing Touches

Two important items shouldn't be overlooked. First, when a deep cut or fill is encountered, the surveyor's stakes must be raised or lowered with the fill or cut by the grade setter and the grades adjusted accordingly. If a cut is no more than about 2 feet, the equipment can work around the stake, leaving it on a mound until the rough grading has been completed. Then the stake is reset and the mound is wiped out. If it's the last pass in the street, the mounds can be cut and the stakes not reset. Second, when the curb is poured, you must mark the location of water services and sewer services on the curb so they can be located later. If there are no curbs, mark the locations with a 2 x 4 redwood stake.

Clean-up follows once the subdivision roads have been paved. A grader should dress up dirt behind the curbs or walks and also any rough areas left at the front and back of the lots. All stakes and debris left during construction must be hauled off. All the manholes, water valves and sewer clean-outs in the streets and easements should be raised. Measure and mark these on the curb before covering them during the subgrade operation. After the manholes, water valves, and sewer clean-outs have been poured and paved around, the subdivision should be complete and ready for the final inspection.

Excavating Commercial Sites

Excavating for commercial buildings and apartment houses is a little different from most of the other work you're likely to handle. There are three major differences: excavation for parking areas, landscape areas, and the swales between buildings. For example, commercial sites will always have at least one parking area adjacent to the building pad. The grading contractor will usually excavate these parking and swale areas.

Take Time for Planning

Take time to study the job specifications and soils engineers' report before work begins. Here are some of the things to look for: Do you have the contract to fill all the planters with topsoil? How much contour grading is expected? Is there any topsoil to be stripped and stockpiled before excavation begins? How much are the pads to be overbuilt and what compaction is required?

Next, arrange a job conference to share information and plan the work. The foreman, superintendent, and estimator should discuss how the job was bid. Let's look at a typical situation. Suppose the plans show a 6-inch parking section (aggregate base plus paving thickness) interrupted by landscaped islands surrounded by a concrete curb. Do you grade and pave the entire area and then cut out the islands? Or is it better to work around the islands when grading, laying down the base and paving? Some contractors would excavate through all the interior islands if there's no grade change between the back and front side of the island. Many estimators feel that it's cost effective to rock and pave through the islands and then to cut them out after the paving is done. Grading, installing aggregate trim, and paving go faster if there are no islands to maneuver around.

Before bidding the job, the estimator should decide if the time saved working through the islands justifies the cost of the extra materials used and the time needed to cut and remove paving material from each island. This method is an option only when the island curbs will be placed on the asphalt paving. If the islands are 14 or 18 inch barrier curbs that sit on or below subgrade, they must be placed before paving.

Excavating an Apartment or Office Complex

The first step in excavating an apartment or office complex is usually stripping the grass or brush. The soils report should tell how to handle this. When stripping begins, the foreman should find the excavation pattern that minimizes travel time for the equipment. Usually debris is hauled off or placed at the edge of larger landscape areas.

The second step is removing the topsoil to a stockpile. The soils report will probably tell you what depth of topsoil to remove.

Setting the Boots

While the stripping and topsoil removal is going on, the grade setter should mark all the stakes, or *set his boots* as it's called. He'll drive a lath next to the surveyor's information lath, which usually shows the top of curb or finished asphalt cuts or fills. Then he'll place a mark on the lath (boot). This mark should be 3, 4, or 5 feet above the finished grade. A 2-foot boot is hard to shoot because it's so low. Unless the grade setter is tall, a 6-foot boot is too high to shoot with an eye level. It may take two laths nailed or glued together to get enough height for the grade mark.

The grade setter should set boots based on the finished grade, then add the curb or road section subgrade depth when checking grade from these boots. After placing each grade mark, mark the cut needed to reach subgrade on all the cut stakes. That way the equipment operators can see what cut is needed directly, without having to add the curb height and road section to the surveyor's cuts.

Figure 8-1 shows the lath set by the grade setter. Then the grade setter marks the height of all subgrade fills on the lath using the finished grade symbol and arrow.

Excavation Begins

Once the grade setter has finished marking the stakes, excavation can begin. If possible, do the stripping in the fill areas first. That way the compactor and water truck can get started working in this area before the fill is placed. The specifications will usually require that even undisturbed ground to be filled be compacted to a certain depth after topsoil is removed. Other than planning for the convenience of the equipment, there's no set rule on where to start the excavation. Beginning at either a building pad or parking area is fine.

Usually apartment and office pads extend 5 feet beyond the building line in both width and length. This will make the pad overlap into planter and other areas. The pad is the most important part of the job. If it overlaps into the parking area, no problem; the parking area near the pad won't be cut for a while. That's also true for narrow swale areas between buildings. Over-build the pad 5 feet even if it covers some swale area temporarily.

If possible, build the pads first. Apartment and office site pads are so large that pad grading can begin before rough excavation of

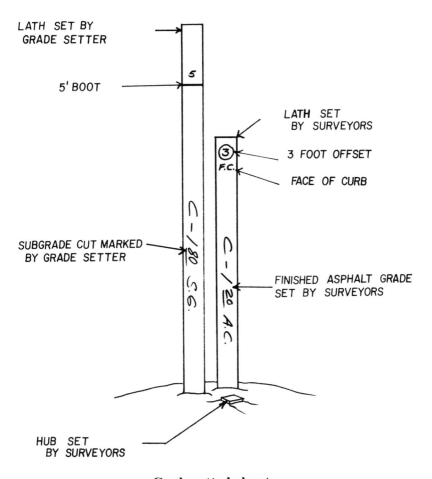

LATH SET BY
GRADE SETTER

5' BOOT

LATH SET
BY SURVEYORS

3 FOOT OFFSET

FACE OF CURB

SUBGRADE CUT MARKED
BY GRADE SETTER

FINISHED ASPHALT GRADE
SET BY SURVEYORS

HUB SET
BY SURVEYORS

Grade setter's boot
Figure 8-1

the site is finished. Then, if more dirt is needed or excess must be removed, the scrapers are available to make quick adjustments. Later a small amount of trim can be pushed into planter areas by the grader.

If there are ramps to excavate for loading docks, be sure to excavate to the back of the concrete structures, *not* to the front when vertical cuts are to be made. Don't leave a slope of less than approximately 2:1 from the top of the building pad down to the planters or parking areas. This provides access to the pad for both equipment and materials.

Excavating Islands

The most troublesome thing about excavating apartment or office complexes is the islands. For large islands, if only a light cut is indicated through an island area and the existing ground is below the top of curb enough for topsoil, you don't need to excavate. If a large cut or fill is marked in an island area, make the cut or fill to the level needed in the island for topsoil placement later.

The grade setter will probably have to remove survey stakes so the scrapers can excavate through islands.

Generally, small islands are ignored. They are cut to parking lot subgrade initially and built up later. The grade setter must offset each stake he removes, then reset it after the area is excavated or filled to the desired grade. There are a variety of methods for doing this, depending on the soil and what's included in the grading contract.

Balancing the Site

If soil in the area will be used in planters and landscape mounds, a small dozer or loader can push or dump the soil into small islands. Use scrapers to fill the larger landscape areas. On most sites, the planters and mounds are left low so foundation dirt can be placed in them later. If a site is balanced and no dirt will be hauled in or out, keep a close watch on the grades. If it looks like there will be too much or not enough dirt to reach grade in all areas, contact the engineer. Let the engineer make the grade changes while the equipment is still working. Otherwise you may have to stop work while corrections are being made. Pads should be overbuilt 0.10 for trim excess.

Remember that some soil will be displaced during construction. Some space is needed to dispose of this dirt to save the cost of hauling it away. Many excavation contractors undercut the parking area enough so it can handle excess dirt from underground spoil. Leave room in the planters and mound areas so there's room to dispose of the foundation dirt. Usually you won't be responsible for disposing of the foundation dirt.

When the initial excavation is completed, have the surveyors check the pads and parking areas to certify that they're excavated properly. That way any excess dirt problems that develop later won't be your responsibility.

The initial surveyor's grade stakes may need to be offset or pulled out and then reset after that area is cut or filled to grade. It's important to leave these stakes in place long enough to keep control of distances and grades. The job will be restaked for fine grading and curbs after the underground work and foundations are in.

After the rough grading is done, you might not be called back until all the buildings are up. When you return to the site, all the underground drains, sewer, and water should be in. If electrical and irrigation lines aren't placed yet, you'll have to coordinate the work of these contractors with your grading to avoid any conflict.

Curbs and Paving

Now the site is prepared for curbs and paving. If extruded curb (curb laid on top of the pavement) is called for, you may decide to grade and pave through the smaller islands.

Be very careful to check the survey stakes and the offsets given. Surveyors follow several different conventions when marking their stakes. Every firm is different. For example, the offset given may be to the top back of curb, face of curb finished asphalt, or edge of pavement finished grade. If it's to top back of curb, the grade setter needs to look at the plan and find the distance from top of curb to finished asphalt. Then he will add that figure to his grades, plus the parking section of aggregate and asphalt for his subgrade elevation.

On the other two markings (face of curb finished asphalt or edge of pavement finished grade), only the parking section of aggregate and asphalt is figured into the grades. Then you decide how far behind the curb to excavate on all the offsets. On a barrier curb, it's good practice to cut 2 feet behind to give the curb crew enough room for setting forms or for the curb machine to work. For extruded curb poured on top of the pavement, leave enough space behind the curb so the pavement can be placed. Also leave enough space behind the curb for the curb machine to work. Usually 4 inches is enough.

Do some planning before cutting grade where curbs will be placed. Some concrete curb machines need more space behind the curb for machine overhang and string line. In that case, the soil beyond the finished asphalt surface should be cut back. But don't go too far. Don't waste more aggregate and asphalt than necessary.

Once you've decided on the distance to overcut, there's just one more decision: Will the subgrade be processed, compacted and fine trimmed now or after the curb is placed? If the curb will be placed on the finished asphalt, that's easy. The subgrade must be processed, trimmed, rocked, and paved first.

If a barrier curb is placed, the depth of the curb and the section of aggregate and pavement will determine how to trim the upper portion of the grade. If the bottom of the curb and the subgrade of the parking section are the same elevation, you may choose to rough-trim the subgrade and trim the curb grade. Once the curb is poured and electrical and sprinkler lines placed, send the crew back to compact and fine trim the subgrade.

The electrical and irrigation contractors should place their lines deep enough so ripping and processing won't damage them. It's much faster to process and compact subgrade before the curbs are placed. Even if the bottoms of the curbs don't match subgrade, the entire area can still be compacted and trimmed all at one time. After the curbs are placed, you'll only have to trim or fill slightly along the curbs before aggregate is placed.

Don't Forget These Items

If compaction tests are required under the curb, it's much more cost effective to compact and trim the entire area at once. Once the curbs are placed, all the sewer cleanouts, water valves and manholes under the paved section should be measured to various objects so you can find them after they're paved over. Be sure any object you measure to will still be there when you need to measure from it again to locate the valves and manholes after paving. This process is called "tying them out." All drain inlets and manholes with slotted covers used as drain inlets must be set to grade before paving so they can be paved around.

Any area that has unsuitable soil, such as soft or muddy soil or pockets of vegetation, must be brought to the attention of the soils engineer. He will determine how much soil has to be excavated and the material you should use to refill the excavated area. Areas in the building pad with unsuitable soil must be excavated before any pad fill is placed. He may feel that any unsuitable soil in the parking lot or street will dry before the grading contractor returns for final grading. If soft dirt goes so deep that it hinders excavating or placement of underground utilities, it should be removed during the rough grading operation. Excavation of unsuitable material is usually charged under the contract as extra work.

It's important that the foreman keep close track of time spent and material moved or used when unsuitable soil is found. But note that unsuitable soil in landscape or planter areas is usually left undisturbed because it poses no problem unless there's a concrete walkway to be placed over it.

Special Care Areas

You'll nearly always excavate for swale areas around building pads for apartments and offices. Make sure that all swale areas that aren't overlapped by building pads are excavated early. You don't want to generate a lot of excess dirt when the final landscape grading is done. Pay special attention to this potential problem and to all excavation for swales, planters, pool areas and sidewalk subgrade around buildings.

Excavation is always more difficult when the working area is limited. That's a common problem when apartment and office sites are designed for a higher density. Where land values are high and buildings are close together, there may not be enough space to stake the swales between buildings. These are important areas and should be excavated accurately during rough excavation.

If they're not staked, the grade setter should check the plan for the elevation difference between the swale and the two building pads it runs between. If necessary, use the pad stake elevation for control of the swale cut. These areas may be too narrow for a scraper when the buildings are up.

If there's a large amount of stockpiled topsoil to be placed in planters, place it after the curbs are in but before fine trimming for aggregate is done. If the planters are filled after the project is paved, it may cause some asphalt damage or tracking on the new pavement.

If the engineer or building superintendent makes any grade or design changes while the project is under construction, be sure the changes are in writing and signed. This is important, even for minor changes where no extra charge is involved to make the change. Otherwise you're accepting responsibility for the change without anyone's authority. That's never a happy prospect.

Planning Excavation

In this chapter we'll look at the most practical excavation methods and procedures for equipment operators. In general, you'll be most efficient if you imagine that the dirt must be moved with a stick. You must scrape the dirt from each high area to the closest low area without carrying it. Follow this simple rule whenever possible to establish the most efficient method for moving dirt.

The prime objective in planning excavation work is to move the dirt the shortest distance possible from cut to fill. Your plan is efficient if all the dirt on a project is moved from cut to fill without traveling over a previously cut or filled area. This isn't always possible, but it should be your goal.

Planning a Road Job Excavation

Assume you have the road job shown in profile in Figure 9-1. Your job is to find the most efficient excavation plan. The job quantities

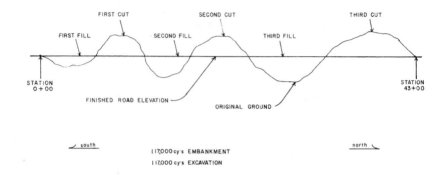

Cut and fill profile
Figure 9-1

are 117,000 cubic yards of fill (embankment) and 117,000 cubic yards of cut (excavation). This is a balanced job because all the excavation will be used as embankment with no excess to remove. You decide to begin the dirt spread on the south end of the job (the left side of Figure 9-1) because that's where the smallest fill areas are located. When the two smaller fills are made, the first cut area will also have been cut to rough subgrade.

Subgrade is the portion of road bed on which pavement base and subbase are placed. Rough subgrade is the grade level cut using crows feet. It's not within the tolerance limit for finished subgrade. Finished subgrade is trimmed level with the grade hubs. The hubs are usually driven in the subgrade to trim to, and are within the tolerance given in the job specifications. Figure 10-1 in the next chapter shows a subgrade diagram for a road cross section.

Next, you usually have the grader start the shoulder trim, cutting ditches and trimming the rough subgrade while the earth moving equipment moves the remainder of the second cut material to the third fill area. When the second cut area reaches subgrade, the scrapers move to the third cut area and finish filling the third fill area to grade. The slopes in the cut areas are cut and trimmed as the excavation proceeds. This finishes the excavating. You'll probably leave only one scraper on the job to move any excess dirt generated by the grader. Fill slopes will be finished with a dozer.

Estimate the amount of excess dirt generated by the grader while cutting ditches through cut areas. Undercut the rough subgrade during excavation so the ditch dirt can be used as fill on the subgrade. This leaves a small amount of trim for the scraper to double back to pick up. If you do this correctly, you can move in a smaller scraper sooner for rough subgrade balancing. This will reduce the job cost.

A choker is a dirt shoulder built up higher than subgrade level. Figure 10-8 in Chapter 10 shows a choker. If you build chokers in the fill areas, leave the rough subgrade low enough to use the excess dirt when the chokers are trimmed. The scraper operators should always leave their cuts as level as possible.

Scraper Techniques

If a push cat is used, set the scraper in position for the next cut close to the dozer. Then the dozer doesn't have to travel too far. Using hand signals, a good dozer operator can direct the scrapers to the area he wants cut next.

If the starting place is the first cut area in Figure 9-1, the scrapers head south while cutting and the cut starts at the top of the hill. The operator cuts as deep as possible at first. Starting down the hill, he raises the bowl slightly. Doing this on each cut will eventually flatten the top of the hill, giving the operator a good flat surface to work from. Then a smooth light cut on a level plain makes loading easier.

Never load going uphill if at all possible. It's very difficult, and it increases the loading cycle time. Figure 9-2 shows a scraper which has leveled the top of a hill after several passes. Now it can load quickly. Notice that there is no smoke from engine lugging even though it's taking a good cut.

Scraper leveling top of hill
Figure 9-2

Start placing material at the lowest point in the first fill area in Figure 9-1. Bring the fill up from the lowest point in smooth level lifts until the desired grade is reached. The scraper operators are responsible for keeping the fills and cuts level.

Use the Right Equipment for the Job

Use the push cat, ripper cat, and compactor for what they do best: pushing, ripping and compacting. Use the dozer on the compactor to mix the fill that was dumped, to keep the moisture distributed evenly, and to push fill to areas not easily accessible to the scrapers. In Figure 9-3, you see a compactor easily leveling and compacting a smooth layer of dirt just dumped from the scraper. Notice how the fill area is kept level.

Keep the Work Area Smooth and Level

Scraper operators have been known to gouge up the cut area and dump piles in the fill areas, relying on the dozer or compactor to

Compactor leveling dirt dumped from scraper
Figure 9-3

level the piles. The operator who thinks he saves time by this quick gouge-and-pile method is badly mistaken. Taking a smooth cut while loading and spreading the dirt evenly during dumping saves time in the long run. Keeping the cuts and fills smooth and level is easier on the operators and makes production more efficient. It also aids the compaction effort.

Don't spin the tires when loading a scraper. It causes excessive tire wear. After a few passes, notice the sound of the engine and the ground speed just when the tires start to lose traction. When it's about to happen, raise the bowl slightly until the engine strain eases and the speed picks up again. A very skilled paddle-wheel scraper operator can take a light cut in second gear and load faster than in first gear. If loading in first gear is easy, try second gear and time the loading period. If the loading time is faster, try third gear with a lighter cut yet. If that's faster, stay with it. If it's slower, drop back to second gear.

There's no rigid standard operating procedure for any one piece of equipment because soil conditions vary so greatly. Whatever equipment you're using, try to develop a faster, easier way of getting the job done. Before starting any excavation, discuss with your foreman and operators the best travel pattern from excavation to embankment. Tell the operators where the excavation and embankment should start. Point out soft areas or underground utilities. Explain the best method of working once the work begins. Encourage the operators to suggest a better plan of operation if they see one.

Haul Roads

The most important thing you can give your scraper operators is a good, smooth haul road. It's probably the single most important factor in achieving good production. Even the best scraper operator's talent is wasted if he has to haul over a rough road. On long hauls, a grader and water truck should make a pass over the haul road periodically to keep it in good condition. This minimizes the scraper cycle time and reduces the man and equipment hours needed to finish the job.

If the haul road has turns, bank the turns so the scrapers can make the turns at a good speed without turning over or sliding off the road.

On small jobs, you may not have a grader available for haul roads. The scraper operators should note any large bumps that are slowing their travel time, and either fill those areas or cut them down on the next trip over. The time it takes to smooth the road will pay off in extra loads because cycle times will be shorter at higher speeds. But note that the scraper operators must be careful when moving at high speeds, especially if a water truck is watering the haul road to control dust. Wet soil can cause a scraper to skid and overturn or hit an oncoming scraper. Slow the speed a little until the water penetrates.

Tank quickly fills a water truck
Figure 9-4

Hauling Water

Just as important as the dirt haul is the water haul. No matter how well the excavation is planned, if water doesn't reach the fill fast enough, scrapers and compactors won't be able to work at top productivity.

Examine your water needs as soon as you've determined how many scrapers and compactors you'll need. If you'll need several water trucks or if the job covers several miles, plan for two or three water supplies. The type of material excavated determines the amount of water required.

In the rainy season, when soil is still moist from rain, only dust control may be necessary. Late in the season, when the ground is dry several feet down, you may need several water trucks. Consider the ground conditions before planning work for the water trucks.

Consider carefully the source of water. If a pump is needed, provide a good base for the trucks to use when loading. Otherwise water spillage can turn the loading area into a mud hole. If water permits are needed, get them before excavation begins. Whenever

Filling a truck with a stand pipe
Figure 9-5

possible, use a holding tank, as shown in Figure 9-4. This tank fills a 3,700-gallon truck in 2 minutes, 18 seconds, then refills itself. A float in the tank shuts off the water when the tank is full. If you use a holding tank, provide a good base and drainage away from the tank.

Another fast way to supply water trucks is with a stand pipe and quick valve. In Figure 9-5, a water truck is being filled from a stand pipe. A quick valve and hose are hooked to a hydrant. This saves about 2 minutes hook-up time per load. Figure 9-6 shows a quick valve. Turning the handle 90 degrees in either direction releases a full flow of water.

Quick valve
Figure 9-6

Highway Grading

This chapter starts with staking a road job and proceeds through the steps needed to complete the job. It also describes excavating procedures. Problems that crop up may change the order of work slightly from what's shown here. Compaction, subgrade preparation and rock grading are covered in separate chapters.

Highway construction is one of the most difficult excavation jobs because of the many changes in grade that occur between the right-of-way line and the centerline of the road, island, or median. In most cases, there's no concrete curb to aid the final grading.

Beginning the Job

First, walk the job with the grade setter to look over the plans and staking before any excavation begins. Place or order all construction signs and barricades. Temporary striping or traffic detours may be needed.

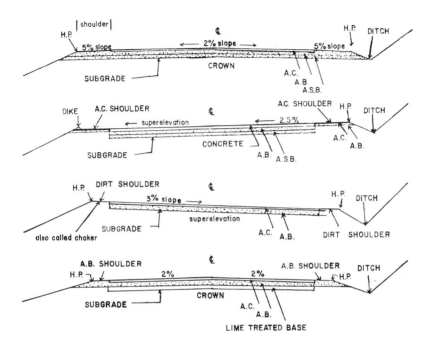

Road cross sections
Figure 10-1

When working adjacent to traffic, be sure all safety equipment is in place before any construction begins. Then check for any necessary de-watering. Determine the number of water trucks required and set up any pumps, stand pipe, or water tanks they'll use.

The first cuts will probably be for a ditch at the top of the slope, followed by a slope cut to another ditch. From the bottom ditch, there may be a fill to the shoulder grade and then a slope to the centerline of the road. If there's an island or median between lanes, those grades will usually be shown on the information stakes. Study the typical road cross sections in Figure 10-1.

These grade changes are less imposing if the men doing the work visualize how the finished grade will look. This should ease any confusion caused by all the stakes the grade setter must set to lay

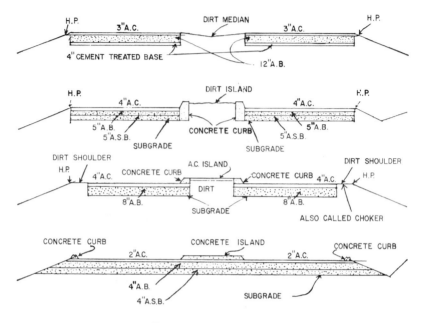

Road cross sections
Figure 10-1 (Continued)

out the work. The foreman and grade setter should watch the excavation equipment carefully on highway work until the operators become familiar with the fill and cut areas.

Staking a Highway Job

When staking out a highway job, the surveyors usually run a row of information stakes and hubs at the right-of-way line on each side of the road. The right-of-way line on each side of the road is usually the limit of the construction area. No work should be done and no equipment should travel beyond the right-of-way stakes without a property owner's permission in writing. If the grade stakes aren't on the right-of-way line, the right-of-way line will usually be staked separately at 100-foot intervals.

Stake Placement

On the stakes at each side of the road are written grades for every change in elevation from the stake to the centerline of the road. Set these stakes parallel to the road, at intervals of 50 feet or less. The cuts and fills start from the given RS point grade. If the road is wide and has a median down the center, you'll probably need a row of stakes there also. If the ground level at the RS point was lowered during clearing, start the fill closer to the information stake to compensate for the cut that was made.

Let's assume you have a fill with a 2:1 slope starting from the RS point given on the stake. When the grass was cleared, 1 foot of dirt was removed, leaving the RS point 1 foot lower. Compensate for this by setting a stake 2 feet back of the RS point on the information stake. At a 2:1 slope, this yields the grade and distance given by the surveyors when the fill is 1 foot higher at the RS point.

Surveyors' Stakes

The surveyor lists all the cuts, fills and distances needed to build the road to the centerline on the front of the survey stake. The stake on the opposite side of the road tells the information needed for the remaining half of the road. The back of the stake gives the station number where the stake is set and the distance from the hub to the center of the road. The distance from RS to the centerline is on the front of the stake. The sides of the stakes show the percentage the road slopes from the centerline to the *HP* or the shoulder, the elevation above sea level at the hub, and the fill or cut slope, such as 2:1 or 1:1.

If the road grade from the centerline to the shoulder is rising at 2 percent, the stakes will read + 2%. If the road slopes down from the centerline to the shoulder of the road, the stake will read -2%. Look back to Figures 1-1 and 1-3 in Chapter 1. In some cases, the survey stakes won't have all this information. It depends on the surveyors and local practice. When the grade stakes don't have all the percentages or information you need, go to the plans for the remaining information. If you can read the grade stakes, you should have little trouble completing the necessary grading as described in the remainder of this chapter.

Setting the Cut and Fill Stakes

First clear all the vegetation, trees and debris from the road area and dispose of it as the specifications provide. Using the grades and distances supplied by the surveyors' information stakes, the grade setter sets out his stakes. Then the operators know where the cut and fill areas are without stopping to read the surveyor's stakes.

The grade setter places the cut stakes first. This means that he puts a stake where the slope starts downward to the road grade. He writes the cut to be made, the rate of the slope, and the distance to the bottom of the cut slope on this stake. Some grade setters indicate only the vertical feet of cut and the horizontal distance to the cut bottom. If a grade setter has experienced operators on the equipment, he can give them the rate of slope and they will know how much to move out with each pass.

Then the grade setter moves to the fill area and sets a stake at the bottom outside edge of the fill to be built up. This is the *toe of slope* and it's one of the given RS points. On the stake at the toe, he puts the same information as on the cut slope: the feet of fill, the distance to the top of the fill, and the rate of slope. The stakes set by the grade setter are crows feet with no hubs needed. Refer back to Figure 4-2 in Chapter 4. If there are any ditches to be cut at the top of the cut slopes or bottom of the fill slopes, these must be staked and cut before cutting or filling the main slope.

Beginning the Earthwork

Once all the outer ditches are cut and the job stripped of all vegetation, the main earthwork can begin. The equipment operators must know the rate of slope of any fill or cut being made. It's very important not to undercut a cut slope or underfill a fill slope.

Restaking Fill Areas

Each time the fill slope rises about 4 feet, the grade setter should set another row of stakes along the full length of the top edge of the fill. Figure 10-2 shows slope stakes set every 8 feet horizontally for every 4 feet of rise on a fill slope. These stakes should have new

Slope stakes
Figure 10-2

fills and distances written on them to indicate what is needed from that point to the top of the fill.

To set a new row of stakes on a fill using a 2:1 slope, the grade setter measures out 8 feet and drives a lath. Then he measures up 4 feet from the previous lath's grade mark and shoots level with an eye level to the lath 8 feet away. At that level point, he draws a horizontal line. The fill must reach this mark.

Setting these fill stakes in the slope periodically helps both the grade setter and equipment operators keep track of the progress of the fill. Figure 10-3 illustrates the markings these progressive fill stakes might carry. If he finds the fill slope is getting too steep, the grade setter can tell the operators to move into the slope more. On the other hand, if the fill is coming up too flat, the operators note the stakes, and fill more to the line marked by the grade setter.

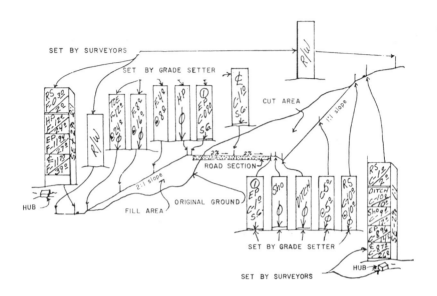

Staking for typical cut and fill section
Figure 10-3

If the grade setter finds an isolated area that needs filling before his stake can be placed, he should direct the compactor operator to push material in and compact it before he sets his stake.

It's good practice to overfill the fill slope slightly so that it's about right when compacted. Usually a 0.50-foot tolerance is allowed on a fill slope. Keep this tolerance on the plus side to avoid problems later when it's more difficult to build the slope up to what's required.

The surveyors should always be notified if the information on one of the stakes doesn't seem to match the others. The surveyors make mistakes occasionally. The sooner they are notified of an error, the sooner your work can be finished.

Cut Areas

In the cut area, keep the slopes shallow enough so the grader can reach them. The cut slopes must be trimmed to grade as the cut proceeds downward. If a grader is used, the cut can be taken down

**First slope pass
Figure 10-4**

**Area after two sloping passes
Figure 10-5**

Grader on a 3:1 slope
Figure 10-6

about 8 feet before it must be trimmed by the grader. Figure 10-4 shows a grader making the first slope pass to cut an 8 foot bank to a 1.5:1 slope. A smooth grade is established at the bottom of the bank so the slope can be cut true. Figure 10-5 shows the area after two sloping passes. The grader now has its front wheels on level ground and the blade is tipped up so the toe can be cut. The grade setter pictured is checking for line at the top of the slope.

Figure 10-6 shows another slope operation. Here, on a 3:1 slope, the grader operator runs the front wheels on the slope to reach the top. By articulating the grader, the back wheels remain on a level surface for a true cut.

Checking the Grade

The grade setter should check the grade ahead of the grader operator and let him know how much to trim. On a 1:1 slope, he'll

Slope stake showing an undercut
Figure 10-7

measure from the top of the slope or previous stake out 5 feet horizontally. Holding the ruler horizontal, he'll use a plumb bob on large slopes. On small slopes, it's easier to drop a pebble from the 5 foot mark. At the point the plumb bob or pebble hits the slope below, set the ruler vertical and shoot level back to the top of slope or previous stake. If the ruler reading at that point is less than 5 feet, more cut is needed. If it's more than 5 feet, too much has been trimmed off.

Slopes

Most jobs use a tolerance of 0.50 foot on slopes. When the slope has been trimmed to grade, the grade setter sets the stakes on the slope with a line indicating *at grade*. He also indicates the cut distance remaining. Figure 10-7 shows a slope stake indicating that the slope has been slightly undercut. A cut of 25.9 feet remains at a 1.25:1 slope from the horizontal line. The horizontal line has an arrow drawn to it.

This procedure happens over and over until the cut reaches the toe of the slope called for on the surveyors' grade stake. The grade setter doesn't need to take his grades from the original survey stake each time. He can take the grade from his slope stakes and change the elevation and distance accordingly but he must be accurate. See Figure 10-3.

Sometimes, in the cut area, there's a ditch on each side of the road that must be cut lower than the finished shoulder of the road. In this case, make the initial cut to the finished shoulder grade. After the dirt has been removed to this point, cut all the lower grade elevations such as ditches.

Chokers

If there are shoulders which should remain higher than the subgrade (usually called *chokers*), this slope or vertical cut should be trimmed. The choker must be cut on a straight even line. If it's overcut, excess base material will be wasted to fill the overcut choker. That's expensive. In Figure 10-8 you see a 2-foot vertical choker that's been compacted and trimmed on a neat line. The subgrade has been compacted and trimmed and is ready for aggregate. Notice that the bank plugs for grade control have been set.

Remember this: It's the finish grade the surveyors are staking. The road section thickness must be subtracted when staking subgrade for cuts and fills. Leave dirt shoulders or chokers 0.10 or 0.20 foot high so they'll be easier to finish after the road has been paved.

Many of the roads you work on will have dirt shoulders and the shoulder finish grade will be higher than the road subgrade. Once the highest grade has been reached in a cut area, move the equipment to the inside of the shoulders and continue cutting until the road subgrade is reached.

Bank plugs set for grade control
Figure 10-8

In fill areas, bring the fill up to the road subgrade and then build up and trim the shoulders to the width specified. Leave the road subgrade low enough so that when the shoulders are trimmed the excess dirt brings the subgrade up to the correct grade. This eliminates extra trimming work on the subgrade. If the road has a dirt-filled island or median, handle it just like a shoulder or choker.

The grade setter must be alert when the cut operation is approaching the shoulder or island grade and the fill operation is nearing the subgrade elevation. The operators need plenty of stakes and guidance so they don't cut or fill too much or in the wrong areas. Make sure each operator has a good picture in his mind of what he's building. This may save the grader setter and the foreman a lot of trouble.

The fill slopes must usually be rolled and trimmed once they're up to grade. You can compact the fill slope while the fill is coming

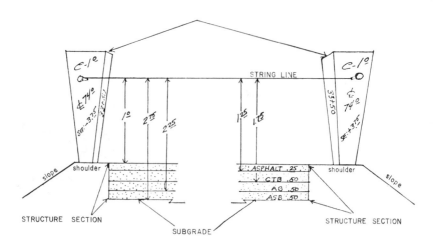

Bank plugs
Figure 10-9

up if a cat and sheepsfoot roller are available. Have the operator run the sheepsfoot down the fill each time about 8 feet of fill has been deposited. This way, the sheepsfoot can be used on the slope while the dozer remains on the top of the fill. If a self-propelled compactor is used while the fill is built up, a sheepsfoot roller may be required later. Use a sheepsfoot roller attached to a cable and crane if the fill is too high for a cat-pulled sheepsfoot to reach down or push up. Often, track rolling with a dozer is all that's required. A dozer can then trim the slope to the required grade.

If a more uniform, smoother grade is required, drag a large chain or cable that's hooked to a dozer or grader from the top and weighted at the bottom along the slope. It'll remove all equipment tracks and lumps.

Subgrade Work

The road is ready for subgrade work when the slopes have been compacted and trimmed and the shoulders are built and cut off vertically on the road side. The surveyors may then set bank plugs for the final grading operation. See Figures 10-8 and 10-9. If bank

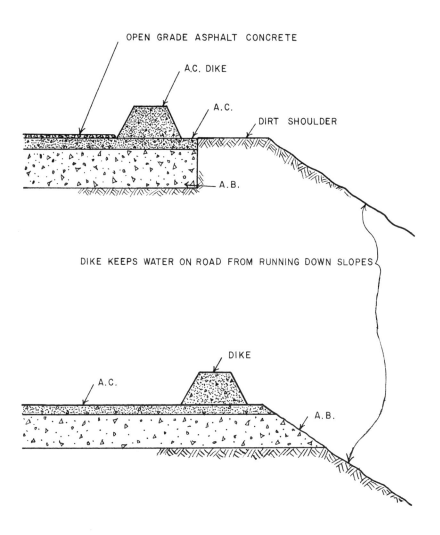

OPEN GRADE ASPHALT CONCRETE

A.C. DIKE

A.C.

DIRT SHOULDER

A.B.

DIKE KEEPS WATER ON ROAD FROM RUNNING DOWN SLOPES

DIKE

A.C.

A.B.

Dikes
Figure 10-10

plugs are set, no other grades are needed for subgrade, rock, or pavement grade. If the surveyors don't set bank plugs, they'll set grade hubs in the subgrade after it's been rough graded and once again when the aggregate base is ready for fine grading.

Trim the subgrade of the road to approximately 0.10 foot high before beginning ripping and compacting. After compaction, trim the subgrade again, this time to the tolerance in the job specifications, usually 0.05 or 0.08 foot plus or minus. Be sure the entire width needed for base or pavement has been excavated and compacted. Never let chokers encroach into the base area.

If the subgrade passes the compaction tests, you're ready to place road rock and pave. The thickness of the rock and type and thickness of pavement to be placed vary greatly. Later chapters cover this work. When paving is complete a final fine grading is usually required.

Keep the shoulders, islands, and ditches trimmed of any excess dirt. Roll the shoulders and shape the ditches. The road can then be washed. Now's the time to place any necessary dikes.

If an open grade asphalt is called for on the top lift (road surface), be sure the dikes are in place before the open grade asphalt is applied. Open grade asphaltic concrete (sometimes referred to as popcorn) is coarse asphalt concrete, usually having 3/4-inch aggregate with very little fine material added. If a dike like the one in Figure 10-10 is placed on it, water will seep under the dike and out of the back side. The dike is intended to prevent water from running off the road and down the slope. If you place the dike before open grade asphalt is put down, it'll collect water as it runs off the road and carry it to a drain.

And now the final steps. Place road signs, paddles, and a fog seal of oil if required. Road striping finishes the job. Finally, take down the construction signs and remove the barricades. The job is done.

Widening Rural Roads

Sooner or later you'll get a contract to widen an existing road. That means driveways, drains, mailboxes, trees, and signs to remove or relocate. For most road widenings, you add 6 or 12 feet on each side. Then the new sections are brought up to the existing road elevation. Finally, the entire road surface, new and old, is overlaid with 0.17 foot of asphalt.

Minimize the Inconvenience in Residential Areas

Road widening is most common in residential areas. And unfortunately, working in a residential area is always difficult. Good public relations are the key to making it work. Talking to the residents before starting may save you some time and money. Explain to the residents that there may be dust, noise or other problems. Tell them that you'll do everything possible to minimize these problems and ask for their understanding. Be courteous to anyone who complains about the inconvenience. And follow through on your promises.

Keep the road free of dust. Wash or sweep it at the end of each shift. Use extra barricades to assure the safety of the residents. And be just as concerned with the safety of your crew working adjacent to the traffic. Make sure that flagmen and delineators provide for safe traffic flow through the project as the job proceeds.

Some mornings you may find flooded areas where runoff from lawn sprinklers has accumulated. Most people will cooperate if you explain the damage and inconvenience they can cause by allowing surface water to run into the cut and fill areas.

Preparing the Area

The surveyors set a line of grade stakes down each side of the road, usually at the right-of-way line. Each stake gives the information needed to build the road section to the centerline. Once the job has been staked, the clearing can begin.

The first problem on a road widening job is the existing utilities. Power poles, gas lines and water lines must be removed or lowered so they won't interfere with the new construction.

Manholes, water valve boxes, or anything in the street that will be paved over must be "tied out" so they can be uncovered and raised later. The best method is to record the station number location of the object being tied out. Then, to be doubly sure you can locate it later, measure its location from two substantial objects such as trees, power poles, or fence corners. Take the measurement from each stationary point to the point being tied out. Enter the distance under the station number already noted. As a final precaution, make a drawing too. It's a good practice to mark two stationary objects with a spot of paint. When it's time to relocate the object that's tied out, measure from the paint spot. Marking with a spot of paint insures that precise spot will be used each time the measurement is taken.

Remove all the home mailboxes and reset them with dirt, sand or gravel in 5-gallon buckets. Set the buckets back out of the way. Relocate signs where necessary but keep them standing during the work. Stop signs, speed limit signs, and other traffic signs should have wooden bases attached so they remain standing but portable. Place a few sand bags on the sign stands to keep them from blowing over.

Next, remove grass, brush, fences and trees. Any asphalt or concrete that will be removed must be cut or sawed at the removal line. Then use a backhoe or loader with a 4-in-1 bucket to start removing asphalt or concrete driveways, walks and drainage pipes.

Try to keep existing drains and ditches open as long as possible. Then if a water main is hit, the water won't flood the yards and homes which still have usable drains.

After all the clearing has been completed, including ditch water and mud, the dirt excavating can begin.

Dealing with the Traffic

The main problem in road widening work will be the traffic. Usually, all the excess dirt must be pulled onto the existing road shoulder. A small paddle-wheel scraper or loader can pick it up there. You need to use part of the road surface for this work.

On two-lane roads, station a flagman at each end of the work to limit traffic to just one direction at a time. That gives you at least one lane for your equipment. The flagmen can hold up the traffic at both ends while the equipment temporarily blocks the road.

Close driveways for the shortest time possible. Notify the property owners before obstructing the driveway, letting them know about how long the driveway will be closed. They may want to get the cars out of the driveway before construction starts. When you put in a driveway culvert, be sure that all the material required is on hand before obstructing the driveway. As soon as the ditch is cut to grade, lay the culvert and backfill so the driveway can be reopened.

The Excavation

The back slopes and ditches will be your first excavation work. If there are areas where the ditch bottom is below the grader's reach, you'll need a backhoe.

Dozer and sheepsfoot roller
Figure 11-1

If getting aggregates to the job by truck at a specific time is a problem, consider renting a vacant lot where aggregate can be dumped and hauled by scraper for pipe backfill or a small fill that might be needed quickly.

Use any dirt excavated from the ditch that's free of vegetation in the fill areas. A small dozer pulling a vibratory sheepsfoot or pad drum roller is a good combination for compacting this or any fill on a narrow shoulder. See Figure 11-1. The dozer mixes water into the fill and the sheepsfoot compacts to the required 95% compaction.

Figure 11-2 shows a pad drum vibratory roller. It works well on small fills (8 feet or wider) and when a large amount of mixing isn't needed. It's also good for initial subgrade compaction when it's followed with a smooth drum vibratory roller. After the back slopes and ditches have been trimmed, the road structure section

Pad drum vibratory roller
Figure 11-2

can be built up. Remember, ditches are a special concern if private yards butt up to the back side of the roadside ditch. In many cases there won't be enough public right-of-way to clean the ditch from the back side. If so, all the excess dirt or aggregate must be pulled up onto the new pavement later.

This is the reason for keeping as much dirt and aggregate as possible from rolling back into the trimmed ditch. If there was a fill slope built during the excavation phase, it may be easier to trim it with a grader or hoe before trimming the subgrade. Experience is very helpful here. If you've worked on similar jobs, you'll remember the problems you faced and can take steps to be sure the same mistakes don't happen again.

Excavating Narrow Road Sections

Here's the best procedure for excavating, compacting, and trimming a narrow road structure section. Assume the new section will

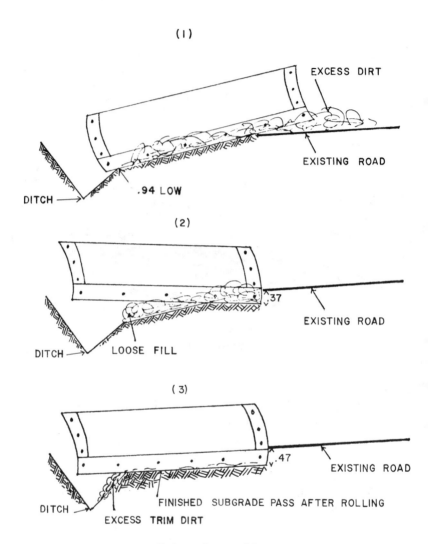

(1)

EXCESS DIRT

EXISTING ROAD

.94 LOW

DITCH

(2)

.37

EXISTING ROAD

DITCH

LOOSE FILL

(3)

.47

EXISTING ROAD

FINISHED SUBGRADE PASS AFTER ROLLING

DITCH

EXCESS TRIM DIRT

Subgrade notching
Figure 11-3

be 0.64 thick, including 0.30 of aggregate base and 0.34 of asphalt. The grade setter has the grader cut 0.94 deep at the outer shoulder edge, with the inside edge held at the top of the existing asphalt. See section 1 in Figure 11-3. When this pass has been made, pick

Vibratory roller
Figure 11-4

up the excess dirt on the street. The section is now ready for the second pass.

On the second pass, the grader operator holds the tip of his blade right at the inside edge of the existing asphalt. He cuts 0.37 at the pavement edge while holding the shoulder edge of his blade high enough so that loose dirt doesn't fall in the ditch. See Figure 11-3, section 2. Next, the section is ripped, watered, and compacted with a small dozer tractor and sheepsfoot roller. After compacting, the grader makes the final passes. The grader operator sets the pavement edge of his blade at the full 0.47 depth. Again he holds the outside edge of his blade high enough so he doesn't lose dirt in the ditch.

A smooth drum vibratory roller should follow the grader on this pass. Figure 11-4 shows a vibratory roller. This roller is excellent for subgrade or aggregate rolling. Notice the section of cyclone fence hooked over the drum as a cleaner.

(4)

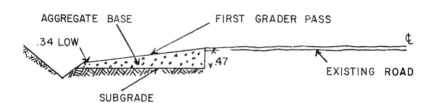

(5)

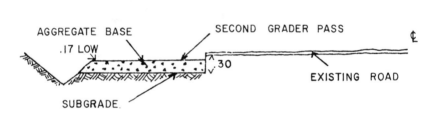

(6)

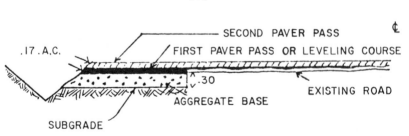

Finished road section
Figure 11-5

The section is now ready for the final trim pass. On this pass the grader operator again sets his blade against and 0.47 below the existing asphalt. But this time he'll hold the outside edge of the fill on grade, letting any small amount of excess dirt roll off into the ditch. This will also add enough width overbuild to the shoulder to keep the aggregate from rolling into the ditch when it's placed. See Figure 11-3, section 3. If you build the section this way, there should be very little dirt falling into the ditch.

Place aggregate base as shown in the top two sections of Figure 11-5. In general, place the base the same way you removed the dirt. And here's a useful tip. When placing aggregate in a narrow area, install wings on the grader blade to control the aggregate. Look at the grader in Figure 11-6. The wing on the end of the mole board controls the aggregate base so excess won't be lost in the shoulder ditch. Notice the hub set to A.B. grade. It has a plastic skirt on it that will pop back up after the grader passes.

Grader blade with wings
Figure 11-6

Trucks dumping the aggregate must run with one wheel in the new section. With the truck bottom dump controlled, portions of aggregate can be dumped by hand right on the edge of the existing asphalt. The grader spreads the aggregate, holding the outside or shoulder edge 0.34 below the existing asphalt. Then it should be rolled. See Figure 11-5, section 1.

On the second pass the grader cuts 0.17 into the lift just spread at the existing asphalt edge. The operator should hold the shoulder edge high enough to keep the aggregate from going into the ditch. This leaves the surface level and 0.17 below the existing road. After rolling, make the third pass holding the edge of the blade 0.17 below the existing road surface and holding the outer shoulder edge to grade. Allow any small aggregate excess to roll down the slope. This is the same procedure used with subgrade shown in section 3 of Figure 11-3.

Paving

Now you can move in the paving machine. If an overlay with a leveling course is required, the job is paved in two passes. The first pass covers the road from the shoulder edge to just short of the centerline. The asphalt is laid 0.17 deep at the shoulder edge and tapered to nothing at the quarter crown just short of the centerline. Look at the bottom section of Figure 11-5.

Figure 11-7 shows a leveling course placed over the old pavement as 0.17 of asphalt paving is placed on the new 8-foot shoulder. Notice the screed dragging on the right side of the existing pavement. Now the road is ready for overlaying from shoulder to shoulder. A final 0.17 of asphalt over the entire road brings the road section to finished grade.

In warm weather, allow the asphalt concrete to cool three days or more. Then pull any excess dirt from the ditches up on the new pavement and haul it away. In hot climates, you may have to pull up the excess ditch dirt in the morning hours only. The sun may soften the new A.C. so much in the afternoon that it will mark

Leveling course placed over old shoulder
Figure 11-7

badly. Then raise, finish and reset the manholes and water valves, pave the driveways, and set the mailboxes. Finally, wash the street and place any required dikes, signs or guard rails. The street can now be fog sealed (if required), striped, and put into service.

Building Narrow Embankments

You'll face special problems when adding a narrow fill section to an existing highway. This is a common situation when widening an existing highway.

Suppose the existing highway must be widened 12 feet and filled 20 feet to meet the existing road surface. Also, assume that the existing slope is at 2:1 and the new fill must be at 2:1. This means that the fill at the bottom must be 12 feet wide plus or minus 6 feet that must be cut into the existing slope to tie in. This *benching* must continue to the top of the fill. Figure 12-1 shows the process of benching into an existing slope.

But here's the catch. A fill area 12 or 18 feet wide doesn't allow enough space for the equipment to pass. If a dozer and sheepsfoot roller are on the fill, then there won't be enough room for the scrapers to pass them.

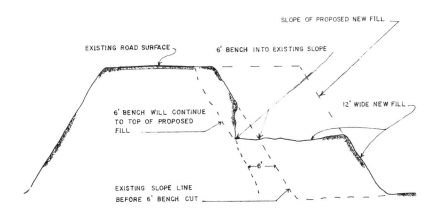

Benching into existing slope
Figure 12-1

Two Ways to Build a Narrow Fill

There are two common ways to build a narrow fill. If you can't close a lane of the highway, you'll have to use the first method.

When You Can't Close a Lane

In the limited space you have to work with, the scrapers can't spread the soil until they've all been loaded. When all the scrapers are loaded, they move into the fill area together, one behind the other. When the dozer operator doing the spreading and compacting sees them coming, he moves the dozer and sheepsfoot or self-propelled compactor out to the end of the fill. He waits there until all the scrapers have dumped and passed through.

If the fill needs water, the water truck follows the scrapers over the fill area before the dozer or compactor returns to the fill. The water truck should try to add most of the water needed in the cut area so few passes are needed in the fill area. But the water truck driver must be careful not to get the cut area so wet that the scrapers lose traction.

This is a slow and expensive operation, but if a lane of the highway can't be closed, it's the only practical method.

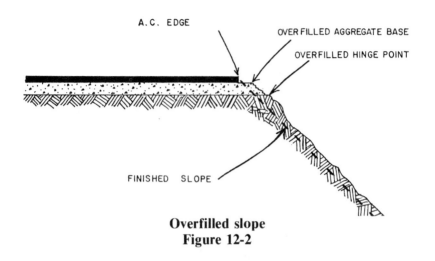

Overfilled slope
Figure 12-2

When You Can Close a Lane

If you can close a lane, you can use the second method of building up a narrow fill. Just dump fill on the closed lane and push it over the side to the dozer and sheepsfoot roller. A model 16 grader can handle about 3,600 cubic yards a shift in this manner. A water truck with a side spray follows the grader, watering the fill when needed. You can use the same operation whether the hauling is done with scrapers or trucks. The grader won't be able to push as much material if it's dumped from scrapers instead of bottom dump trucks.

If it's a deep fill of 25 to 30 feet, with the dirt being bladed over the top, you may need a D-6 dozer on the side of the slope. It depends on the material and percent of slope. A heavy clay will tend to build up at the top of the slope or half way down the slope. You need to give it an extra shove to get it down to the dozer that's pulling the sheepsfoot roller at the bottom of the fill.

Make sure your operators keep the outer edge of the fill higher than the inside. This prevents equipment working on the outer edge from sliding off the fill. Once the fill reaches the subgrade level, work it like any other road section. Be sure to keep the fill 6 inches wider than specified to allow for shrinkage when the slope is compacted. Figure 12-2 shows an overfilled slope.

Excavating to the Subgrade Level

This chapter covers the trimming of subgrade on subdivision streets and highways. It also explains how and when to stake the job and what equipment to use.

On a subdivision job, you're ready to trim the street subgrade when the underground work and curbs have been finished. But before trimming the subgrade, tie out any manholes, water valves or clean outs that must be located and raised after paving. You can do this by measuring from them to points at the back of curbs or walks.

Setting Centerline Grades

The first job is setting the centerline grade hubs. If the curbs are in on both sides of the street, here's the most common way to set the grade. Let's assume the road is 30 feet wide and slopes 2 percent from the centerline to the curb in each direction. That makes the surface at the centerline 0.30 foot higher than the lip of the curb

(2% x 15 feet). If you'll place 4 inches of rock and 2 inches of asphalt, the finished subgrade level at the centerline should be 0.20 foot lower than the lip at each curb. When 6 inches (0.50 foot) of structure section is added, the finished grade at the centerline will be 0.30 foot above the lip of the curbs. This is the 2 percent needed for the crown.

Sweding

Sweding is a fast and very accurate way to set grades. Here's how to do it. Use three laths, two that are 3 feet long and one for the center that's 3.20 feet long. You'll need three people, with one at the centerline driving hubs. Set the two shorter laths on opposite sides of the street on the curb lip, as directly across from each other as possible without measuring from a station. If the curb stakes are still undisturbed, you can use them as reference points. Put the longest lath on the ground at the centerline. Sight with your eye across the tops of the laths. If the center lath is higher than the two side laths, you must lower the grade at the centerline. See Figure 13-1. If for some reason the lip of the curb on one side is a little too high or low, there'll still be an even slope in each direction.

In Figure 13-2 you'll see a grade setter using a swede at the lip of the curb and one at the centerline. He sights over his finger on the ruler at the far lip. Notice his bucket containing lath and hubs. He'll use them to set a centerline grade for trimming.

Using a String Line

Another method uses a string line stretched from curb to curb. The person in the center measures down each time to set the hub. He makes a mark or knot on the string at the center point to eliminate measuring for the road center at each station.

If the grade setter doesn't have an assistant, he'll shoot the grades in with an eye level or laser controlled rod. This third method of establishing the grade takes only one person instead of two or three, but it's the slowest of the three methods. You may

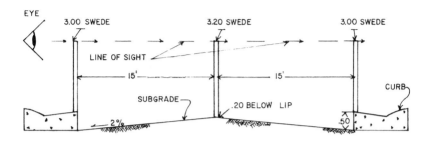

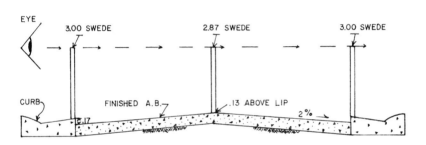

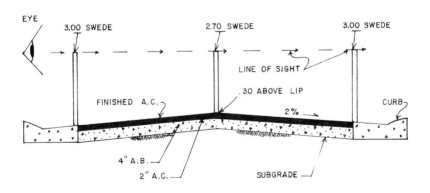

Typical sweding procedure
Figure 13-1

Grade setter using swedes
Figure 13-2

also consider setting lath with clothespins. The method is described in Chapter 4. When setting grade to balance dirt for compaction, set crows feet, not guineas.

Balancing the Subgrade

After the grade has been set, use a grader and paddle wheel scraper
to cut the subgrade to the correct level. This is called *balancing* the
subgrade. It means that all the low and high areas are closely trimmed so they average 0.10 foot high. Leave subgrade soil about 0.10
foot high to allow for shrinkage during compaction. After the
subgrade has been balanced, it should be ripped and watered, mixed to the desired moisture and compacted.

Once the subgrade has been compacted and shaped with the

grader, it's time to set centerline grades again and fine grade the roadbed. In most cases, plus or minus 0.05 foot or 0.08 foot is an allowable tolerance. This operation usually requires a grader, a small paddle wheel, a roller and a water truck, plus a grade setter and laborer.

Cutting Subgrade

Cutting subgrade on highway work is difficult because there are no curbs to work from. On a highway job, the grader setter usually shoots the rough subgrade level from the cut and fill stakes. He runs three rows of hubs; one row on each side or shoulder of the road, and one down the centerline. In some cases, five rows are necessary if the slope at the shoulders is different from the road slope, or if there is an island down the center of the road. If the shoulders aren't very wide, he may elect to carry the road slope through the shoulder area during the rough grading.

There are a variety of shoulder and road designs. The foreman and grade setter should agree on the best method of staking the particular road design before the grade setter starts his staking. Once a method has been agreed upon and the road is staked at the edges and center, the rough trimming can begin. Again, leave the subgrade 0.10 foot high to allow for the shrinkage during compaction. Fine grading is always faster when a small trim is needed throughout the roadway. It takes more time if both trims and fills are needed.

When rough trimming has been finished on a highway job and the subgrade has been ripped and compacted, the surveyors will usually reset the grade before any fine trimming starts. The state, city or county agency decides how the staking will be done. The foreman should consult with the surveyors if he feels that the staking isn't adequate for his needs.

Bank Plugs
A *bank plug* is a wedge-shaped piece of 2" x 4" lumber driven into the ground, with about 24 inches showing above ground level. In most cases, you'll set bank plugs at each side of the road. See

Figure 10-9 in Chapter 10. The surveyors write the required information on the bank plug. They drive a nail in each plug at opposite sides of the street at a given distance above the finished roadway.

If the road has a crown rather than a regular slope in one direction, the surveyors set two nails in the bank plug. To set the grade for one-half the road, stretch a string from the top nail on one side to the bottom nail on the other side. Then rotate the string to the other nails to establish the other half of the road.

If the road has a continuous slope in one direction, it's called a *super.* It'll only need one nail in the bank plug on each side. Stretch a string from one nail to the other and set the hubs by measuring down from the string. It's best to use a good colored fishing line because it's light and strong and won't sag or break when stretched.

In some cases, the super in the road will change. One side may slope 2 percent from east to centerline, and from centerline to the west edge it may drop 3 percent. If so, each bank plug has two nails so the string can be rotated to catch the extra 1 percent fall for the remaining half of the road.

The surveyors will write all the information necessary for finishing the road on the bank plug. It'll show the station number, the percentage of slope of the road, and a plus or minus sign to indicate the direction of the slope from centerline to shoulder. After the bank plugs are set, the grade setter stretches his string line and sets up his grade hubs by measuring down from the string.

Subgrade Hubs
Some engineering firms don't set bank plugs. Instead, they set finish subgrade hubs. In either case, after the grade hubs have been set, the fine trimming can begin and finish in the same manner as on a subdivision street.

A grader operator needs a grade hub set every 50 feet down the road and every 20 feet across the road. If the hubs are set more than 20 feet across the road, it's necessary to check the grade just ahead of the grader as it makes the final pass. Use swedes of equal length to do this. Set one lath on each hub and sight across to the center swede.

An established grade setter usually has his own set of swedes. Normally, each is 1/4 inch by 3½ feet long with a 3/8-inch iron plate for a stand at the bottom. The grade setter may have an adjustable slide bar, or he can clip a clothespin to each swede at any measurement needed, and check the grade unaided. See Figure 13-2.

Once the grade has been checked, he'll write the cut or fill required on the subgrade soil with paint. If the grader is working with the crew sweding the grade, they signal a cut or fill to the operator rather than painting it on the subgrade. Use this method between all hubs that are 50 feet apart down the centerline or on the edge of pavement on roads with no curbs in. This eliminates sags and humps between hubs.

Trimming subgrade in a narrow area may require the grade setter to check grade at each station if laths and hubs will be in the way of the grader. The grade setter must be fast and accurate so he won't hold up the grader. As he shoots grade at each station, he'll signal to the grade operator what is needed at that station. See Figure 13-3.

Grade setter checking subgrade
Figure 13-3

Equipment Techniques

Carefully control the amount of water used while trimming subgrade. Soil that's too wet or too dry can't be trimmed properly. The roller operator must wait until the water soaks in before rolling the subgrade. If he doesn't wait, the roller will pick up the wet soil and leave a rough surface.

Make sure that the top layer of soil isn't dryer than the lower layers. Rolling a dry top layer over a wetter bottom layer causes the top layer to crumble, leaving the surface with a scaly appearance.

For fine trimming, you need a good operator on the paddle wheel scraper. He must be able to pick up the excess dirt without cutting into the trimmed surface. See Figure 13-4. It shows the operator of a self-loading scraper picking up a small windrow of dirt left after the grader has trimmed subgrade.

Picking up a small windrow
Figure 13-4

Don't spend too much time picking out small spots that seem to be a little rough, unless of course the inspector complains. When your grader operator starts working with small areas he'll usually create more problems than he solves and will certainly waste time. A lot of time and money can be wasted trimming subgrade if your operator is too careful or inexperienced. Remember 0.05 or 0.08 is usually the tolerance allowed.

Check Your Equipment

Check the air pressure in the scraper tires before any trimming begins. If one tire has as little as 10 pounds less pressure than the others, the scraper will lean and dig deeper on the soft tire side as the scraper bowl fills. Even an experienced operator has trouble when one tire is soft. The operator must keep the windrow centered with the scraper so one side won't fill faster, and cause the scraper to lean and dig deeper on one edge.

The operators and the foreman should always keep a close watch on the cutting edge of the equipment. It's costly to repair a worn mole board on a blade or the pan of a scraper. Moreover, a piece of equipment with a badly worn cutting edge isn't efficient. For finish work such as subgrade trimming or base rock trimming, a good cutting edge is essential. A badly worn cutting edge is seldom level on the bottom and the ends are nearly always worn. It takes much longer to get the grade to within 0.05 foot with a worn blade. If the cutting edge is bad enough, the blade man won't ever make the cut properly.

Watch the cutting edge on any paddle wheel scraper that is picking up soil in a trim operation. A scraper with a worn cutting edge or with worn slobber bits won't be able to make a clean pass and will leave more work for the grader operator.

Slobber bits are the two pieces of steel on each side of the scraper bowl that keep the material confined to the front of the cutting edge until the paddles can scoop it up. Be sure the slobber bits are extended so they are close to the bottom of the cutting edge, but never lower. Figure 13-5 shows how the cutting edge and slobber bits should look when they are in good condition. They all

Cutting edge and slobber bits
Figure 13-5

touch the ground evenly. The arrows point out the slobber bits on this self-loading scraper.

When working a dirt spread that's in hardpan, be sure the ripper teeth are in good condition on the ripper cats, scrapers, or graders. Teeth that are sliding in the same grooves each time without ripping are almost sure to be dull.

Drainage Channels

In this chapter we'll look at the preferred method for staking and excavating drainage channels. We'll also consider some problems that may arise and the best ways to minimize them.

Controlling the Water

Usually, the main problem on existing channels is keeping the water flow out of the channel while you're doing the work. The method you use for dewatering usually depends on the amount of room available to divert the water. You'll need to build a dam on each end of the channel project. The height of the dam depends on the amount of water to be held.

On a small channel, sand bags and plastic tarps may be sufficient. On large channels you may need to use dozers and scrapers to build a large dam.

Trenching

If there's enough room to detour the water around the work area by trenching or excavating from the upstream side of the upper dam to the downstream side of the lower dam, that's the best method to divert the water. It usually doesn't require maintenance.

Pumping

Pumping may be required when working in an existing easement through the backs of yards where no room is available for rerouting water by trenching or excavating. Gas pumps or a generator with electric submersible pumps can pump to a nearby storm manhole or along the channel through a pipe to the downstream side of the lower dam. A diesel generator with submersible pumps and fuel capacity to run all night works well if you need 24 hour pumping.

If pumps are left unattended all night, build a fence around the area to protect against vandalism. Consider a security service for periodic checks during the night as well. The cost of a security service is minimal compared to the cost of cleanup if the dam overflows because of vandalism.

Protect the pump intakes from debris floating in the water. Put them in 55 gallon drums that have slots cut in them. Water can enter through the slots, but trash is blocked. You can also use a screen around the intakes. Both methods must be maintained to assure no stoppage occurs.

Here's an alternative to consider: Run a pipe deep enough below the channel floor so it won't be damaged by the excavation work. Study the diagram of this method shown in Figure 14-1. You'll just plug the pipe and leave it when the job is finished.

Many problems can be avoided in the beginning if you select the right equipment for the job. Equipment too big for the work or too much equipment results in lower productivity every time. Plan your equipment needs carefully on a channel excavation job.

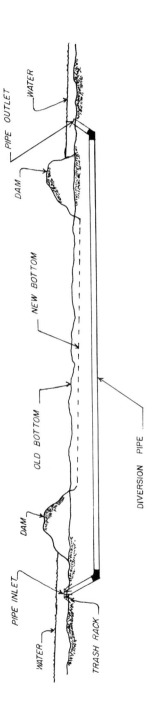

Diverting channel water
Figure14-1

Staking the Channel

A channel is usually staked with two rows of stakes, one on each side of the channel. The stakes should be offset far enough from the top edge of the channel slope so they don't interfere with the equipment. Mark the fills or cuts computed from the hub at each stake for the following distances; the distance out from the stake where the cut begins (top of slope) and the bottom of the cut (toe of slope), the distance to the center of the channel. The stakes are usually set every 50 feet.

Cutting the Channel

When cutting a channel, first determine the best locations for the ramps in and out of the channel. In most small channels there's not enough space for one piece of equipment to pass another. Work out a good route so the equipment returns to the entrance ramp as directly as possible.

Narrow channel work can be very slow because grading and excavating equipment can't work simultaneously. Still, operators must be very careful not to undercut the slopes. When excavation along the slope has reached a depth of about 8 feet, the slope should be trimmed with the grader. On a narrow channel the scrapers may have to wait for the grader to finish the sloping pass before they can get by. If this is the case, it's essential that the grader operator be both accurate and fast so he doesn't hold up the scrapers any longer than necessary. Channel cutting can be very difficult for an inexperienced foreman.

In some cases, you'll need the dirt dug from one section in a preceding section of the channel to build the sides before cutting. Once the channel is dug, remove the ramps. Trim the sides where the ramps were with a hoe or Gradall, or a dozer if the slopes are flat enough and the bottom dry and wide enough.

Widening an Existing Channel

Widening an existing channel can be a real problem. There are several different ways to approach this. The first requirement is to reroute the water around the construction area. If the bottom is muddy and grassy, use a hoe, dozer or dragline to clean it out down to stable material. Then haul dirt in from a borrow area and fill the channel to the top. After that, do the earth moving just as if the channel were given a new alignment. If the channel will be widened enough, start a bench on each side and excavate a new slope with scrapers.

When reconstructing washed-out channel sides and bottoms, the work will be very slow if no borrow dirt is available. That's because excess soil from slope trimming must be used to fill the existing slope and bottom voids. The only advantage in this case is that no equipment will be traveling on the channel bottom, since you'll probably be using a Gradall. So you can leave the bottom soft, unless a concrete lining will be poured.

Trim the slopes with a hoe or Gradall. In areas where slope filling is needed, build a bench or terrace and then overfill the slope so the hoe will only trim and not fill. This is an expensive way to do this work. First the hoe trims the dirt. Then the dirt is hauled to the slope that's being filled. Finally it's recut by the hoe. The dirt is actually moved three times.

As you can see, channel reconstruction can be difficult. Most problems are the result of the lack of working space and the soft bottom in small existing channels through narrow easements. When widening an old channel where the cut on the top isn't wide enough for the scrapers to get started, use a dozer. The dozer starts by dozing the slope out to a point where it's wide enough for the scrapers to begin excavating. This is called *benching*. Use this same method to widen cut slopes on road jobs. When bringing down the cut slope, the dozer operator must be careful not to undercut the slope. If lining the channel with concrete, take care cutting the slopes to keep them as straight as possible.

Cutting in Ground Water

On a large drainage channel where the water table is a problem, you may have to excavate in two stages. Use scrapers for the first stage until the bottom becomes too wet and soft. Trim the slopes of the first stage with a grader. Then a hoe or dragline finishes the remaining excavation to the bottom.

When ground water is a problem, use a well point pumping system to control the water level. This helps in two ways. First, it allows the scrapers to work longer before the hoe or dragline is needed. Second, when the hoe or dragline is finally required, the well points reduce the amount of water in the channel, making excavation and grade checking easier.

You may need planking or sheeting under the tracks of a hoe or dragline to prevent the equipment from getting stuck. If the bottom is firm enough, use a small dozer with mud tracks and a slope bar to trim the slopes and the bottom to grade. This relieves the hoe or dragline of this task and improves production.

There's usually little cleanup left to do when the channel excavation has been completed. You'll only need to reset existing fences or construct new ones and remove the ramps and dewatering system to finish the job.

Curb and Sidewalk Grading

Now we'll look at the the grading steps to follow when cutting the two most common types of curbs, rolled and vertical. These are usually referred to as type 1 and type 2 curbs. See Figure 15-1.

Staking Curbs

All the grade stakes must be set before beginning work on the curb. There should be at least a 2-foot offset from each hub to the back of the curb or sidewalk. The stations set by the surveyors should be a maximum of 50 feet apart. All gutter drains and summits should be indicated. A summit is the highest point on the road or street cross section. Water flows from the summit, in two directions, to low areas where the gutter drains are located.

All corners should have a grade stake at the beginning and the end of the radius, as well as one or more grade stakes between those two points, depending on the size of the radius. There should be a radius point away from the corner so you can use a tape

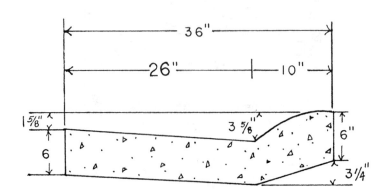

Type (1) Rolled Curb

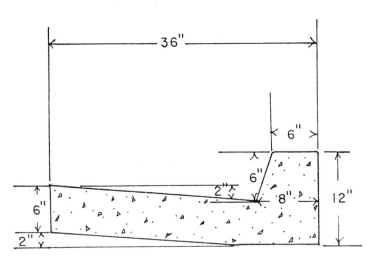

Type (2) Vertical Curb

Rolled and vertical curbs
Figure 15-1

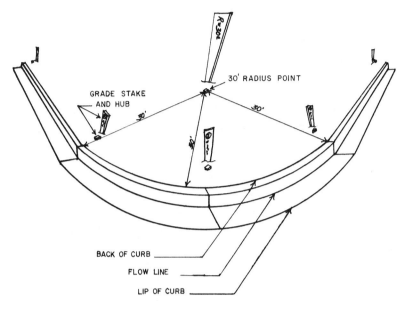

GRADE STAKE
AND HUB

30' RADIUS POINT

30'

30'

BACK OF CURB

FLOW LINE

LIP OF CURB

Radius point staking
Figure 15-2

measure to check the distance to the curb or walk at any point along the radius. Begin the curb grading only after the curb has been staked correctly. Figure 15-2 shows a properly staked curb at a corner.

Type 2 Curbs

Type 2 curb is the easiest curb to grade. The grade setter should check the curb grade with a straight edge and a hand level. This method is faster than shooting with an eye level. The straight edge should have a 1-foot vertical extension on one end and a hand level taped close to the opposite end. If the area in front of the grade stake is rough, have the grader make a pass along the front of the stake line to smooth off the area that will be cut. Once a rough pass has been made, start the first cut. The grade setter keeps checking the ground level against the grade given for the curb by the surveyors and the grader continues making passes until the soil is at the correct level.

Assume that the first stake set by the surveyors reads (2) T.B.C. C-1^{00}. This means that the hub is offset 2 feet and the top back of the curb is 1 foot below the level of the hub. The grade setter measures out 2 feet from the hub and down 3.05 feet from the top of the straight edge to the ground level. This 3.05 feet allows for the 1 foot vertical extension on the straight edge, the 1 foot curb depth, the 1 foot called for on the stake, plus 0.05 for a slight undercut. C-1^{00} + 1^{00} + 1^{00} + .05 = 3.05. Remember that the surveyors always indicate the finished curb level at the top back edge. The flow line of the curb is usually given only for "V" gutters.

Check Figure 15-1 again. Notice that it's 1 foot from the top of the curb to the bottom of the concrete. The grade setter establishes the grade 2 feet out from the hub, but the grader operator cuts to within 1 foot of the hub. This allows 1 foot of working room for the forming crew or the concrete machine. The grade setter holds the end of his ruler on the spot while he signals to the grader operator the amount of cut or fill remaining at that station. He repeats this method at each station until the grade required has been cut. Once the back of the curb grade has been cut, the grade setter checks the grade at the front lip of the curb.

Notice that the grade at the front of the type 2 curb in Figure 15-1 is 2 inches higher than the back of the curb. The grade setter measures out 5 feet from the hub and down 2.90 feet. The 2.90 feet is the result of subtracting 0.17 foot (2 inches) from the 3.05 used to cut the back curb grade. Subtracting 0.17 from 3.05 leaves 2.88. Most grade setters round 2.88 off to 2.90. The front edge should be very close to grade already if the grader operator is experienced enough to hold a slight sloping angle while cutting the back grade. The edge of the blade should pass over the place where the back of the curb will be when the front grade is being trimmed.

After the curb grade has been trimmed, a windrow of dirt should be bladed up close to the front lip of the curb. The curb crew uses this loose material for regrading after the forms have been set.

Cutting Curb Grade

Cutting curb grade for curb forms and for a self-trimming curb machine are nearly the same. But note these exceptions. A 3-foot offset will be needed on the grade stakes because some machines need more than 2 feet for clearance. The curb grade must be cut 0.05 to 0.10 foot high to leave the machine a slight trim. It is important not to undercut curb grade for the machine. This will waste concrete. The curb machine cannot make fills — it only trims. If the machine is trimming an undercut curb, it will push loose dirt into small low areas, leaving it uncompacted. That will cause the curb to fail later. Also, no windrow is needed for grading.

Here are two more things to remember: The machine needs more room behind the curb, and the grade must be pre-ripped and compacted in hardpan areas. On a large job where a pretrimmer will precede the concrete machine, leave the grade 0.10 to 0.30 foot high. See Figure 7-6 in Chapter 7. The pretrim machine can be set 0.05 high so the curb machine has only a very slight trim. This will boost production substantially. It also eliminates the chattering and override that happens in heavy cuts, leaving a good smooth job. Look back to Figure 6-2 in Chapter 6.

Type 1 Curbs

Cutting grade for a type 1 curb is more difficult because of the small 3¼-inch high slope that must be cut. See Figure 15-1 again. All measurements are computed from a 2-foot grade stake offset, just as they were for the vertical curb.

Make the first cut 1 foot behind the curb. This is usually 1 foot from the hub. Notice that Figure 15-1 shows the curb 6 inches thick. The grade setter must add 0.55 foot (6 inches plus a 0.05 foot undercut) to his straight edge reading plus 1 foot for the vertical extension, just as for the type 2 curb. Adding 1.55 foot to the cut indicated on the stake and subtracting it from a fill, gives the level for the top back of the curb subgrade only.

Once this back grade has been trimmed, the flow line grade can be cut. First measure out from the hub 2 feet 10 inches. Notice that Figure 15-1 shows a vertical difference of 3¼ inches from the base of the back of the curb to the point below the flow line 10 inches away. The grade setter converts the 3¼ inches to tenths and hundredths of a foot: He adds 0.27 foot more to the 1.55 foot he measured down when setting grade for the back of the curb. Now make the flow line cut 2 feet 10 inches from the hub. Add 1.80 foot to any cut given and subtract it from any fill given by the surveyors. Adding 0.27 foot to 1.55 foot equals 1.82 foot. The grade setter rounds the 1.82 off to 1.80.

The grader operator now makes his cut with the blade angled up slightly so the front edge of the curb cuts to approximately the correct grade also. To trim flow line grade, he'll hold the tip of the blade at the flow line and raise or lower the blade angle to correspond to the angle at the lip of the curb.

The type 1 curb in Figure 15-1 shows all the distance and elevation changes needed for cutting the grade. Notice there's a line extending from the lip of the curb indicating that the lip is 1⅝ inches lower than top back of the curb. Converting 1⅝ inches to tenths of a foot gives 0.13 foot. Add this to the 6-inch thickness of curb to get a measurement of 0.63 foot from the top of the curb to the bottom of the curb at the lip. Adding the 0.05 foot undercut gives us 0.68 foot. Round 0.68 foot to 0.70. Use this distance to cut the front lip of the curb. Remember to add the 1 foot for the straight edge vertical extension.

If the grade setter is shooting the grade with an eye level, all the figures above would be added to his boot at each station. Chapter 4, Grade Setting, gives instructions on establishing a boot.

Once the front lip grade is cut, the only excavation remaining is the slope between the top of the curb and the flow line. If the material being trimmed is fine dirt, the curb crew creates this slope by pulling a grade bar with a small tractor. In hardpan, the slope from the top of the curb to the flow line must be ripped or cut. It's hard for the grader operator to trim this small slope. The grade stakes obstruct the path of the grader working on the small slope.

Using Curb Shoes

Many excavation companies make up *curb shoes* that can be bolted to the blade. The bottom of the shoe is shaped to match the shape of the curb bottom. The shoe is held on the ground so that the outer edges of the shoe bottom match the back curb and lip grade already trimmed. The shoe then trims the small slope neatly. The shoe saves some time and makes the curb grade more uniform.

If a sidewalk is attached to the curb, the curb shoe leaves a notch of approximately 2 inches at the point where the sidewalk meets the back of the curb. This 2-inch notch is the difference between the 4-inch sidewalk thickness and the 6-inch thickness of the curb where they meet. If a curb shoe isn't used, the slope must be cut back a greater distance to undercut for that notch. This creates a small fill at that point for the grade bar when the fine grading is done. When fine grading, a fill is easier to make than a cut. Remember this when cutting curb or sidewalk grade. Always undercut slightly. Any grade that is too high will hamper the hand grading operation.

Using a Self-Grading Curb Machine

When a self-grading curb machine is used, leave the grade 0.05 to 0.10 foot high so the machine can make a small trim. If doing hand grading, leave a windrow of dirt the laborers can use for fine grading. If a sidewalk is attached to the curb, the sidewalk grade is always cut by the grader before the curb grade. Check the grade stakes. If the sidewalk is attached to the curb, the grade given by the surveyors on the grade stakes may be the top back of the sidewalk rather than the curb. If so, check the standard drawings in the specifications to compute the grade changes from the back of the walk to the top of the curb, the flow line, and the lip of the curb.

The only difference between cutting vertical curb grade for a self-grading curb machine and cutting for a formed curb is that you should leave the grade 0.05 to 0.10 foot high for the curb machine.

When cutting sidewalk and type 1 curb grade for the self-grading curb machine, there are two differences: First, in soft soil the slope at the back of the curb doesn't need to be cut; a vertical cut can be left at the flow line. Second, leave the grade 0.05 to 0.10 foot high rather than undercutting it 0.05 to 0.10 foot. The dirt windrow isn't needed for the hand grading.

If a self-grading curb machine is used and the ground is hard-pan, it should be ripped and recompacted. This keeps the machine from jumping and chattering through the hardpan areas. The result will be a better looking job, higher production, and less wear on the curb machine. Be sure to remove any large rocks that might damage the machine.

Unsuitable
Material

Unsuitable material is any soil or aggregate that has absorbed enough water so that it won't give adequate support for a road section. You'll usually recognize it by a rolling movement of the ground as the equipment runs over it. It's similar to the way a waterbed ripples. A surface that moves this way is said to be *pumping.*

Keep Accurate Cost Records

Any time you remove unsuitable material, keep an accurate record of the quantity of material removed and replaced. Any unsuitable material removed below subgrade level is usually allowed as an extra charge. Be sure to read the specifications before removing any unsuitable material and discuss it with the inspector, engineer, or owner. If the specifications or bid items don't cover unsuitable material, don't remove it until you've reached an agreement on payment. It's unlikely that the project specifications will require the contractor to remove unsuitable material at his own expense.

When you do have to remove unsuitable material at your own cost, it's going to save you money if you can tell unsuitable material from suitable material. Distinguishing the two soil conditions with accuracy takes a great deal of experience. Don't entrust the decision to an inexperienced foreman or superintendent. When in doubt, seek advice from a soils engineer or an experienced superintendent or foreman.

Testing for Unsuitable Soil

Earth with a heavy clay content may move under the weight of rollers as though it were unsuitable material. It may actually be quite stable for most purposes. Here's how to test for unsuitable soil: Take a grader, scraper, or water truck and wheel roll the area in doubt. If the area rolls in front of, or behind the tire, then it's too soft. If no rolling occurs, just a slight settlement from the tire, then it's usually stable enough to support a road section. Don't use a steel drum roller. It won't give a good indication of the soil stability.

Removing Unsuitable Material

When removing unsuitable material, make sure you remove enough to expose good firm soil. If a firm bottom can't be reached, let the inspector or owner make a decision on what should be done next. If 4 feet of unsuitable material has been removed without reaching a stable base, you can recommend the following method. It will bridge the unsuitable area in most cases.

Fill the excavated area with large cobbles or pit run gravel. Push the full 4-foot lift into the depression. Don't do any rolling until the depression is filled to the top. If it's placed and rolled in layers, the gravel will usually continue to roll or pump even at the finished grade level. About 2 feet of 8- to 10-inch cobbles, with 6 inches of base is usually enough to bridge unsuitable areas for cars and light trucks, like a parking lot.

Of course, the inspector, engineer, or property owner should agree on which method to use. If the method doesn't provide a suitable base, you don't want to be held responsible. It's always best if a soils engineer decides what is needed to stabilize the area and what results are adequate.

Excavating the Unsuitable Material

When excavating any road or parking area, try to excavate to the subgrade level before removing the unsuitable material. If the unsuitable soil is so soft that the equipment can't get through, you may have to remove it before the subgrade level is reached. If the areas needing excavation are too soft, too deep or too small for scrapers, use a hoe, track loader, or dozer.

If the unsuitable areas are large enough for scrapers and there's a waste disposal area close by, open bowl scrapers with a push-cat can handle the job. If the unsuitable material area is long, you can use either paddle wheel or open bowl scrapers with a push-cat. A paddle wheel scraper can't cut deep enough to get a full load in a short area. Open bowl scrapers can be used in small areas, however, because they have the ability to take a deeper cut when pushed by a dozer.

During the dirt moving operation, avoid running equipment over unsuitable soil areas. This slows the equipment and cuts down production. If there's no convenient way around unsuitable soil, it may be more practical to just remove it as you meet it. This allows the scrapers to haul over firm ground.

Determining How Much Fill Is Needed

If you're using aggregate to fill the unsuitable areas, and if the unsuitable areas are irregular and hard to measure accurately, you can generally figure one cubic yard of dirt excavated for every two tons of aggregate used. This rule of thumb is accepted by most contractors, inspectors and owners.

Plugging Small Unsuitable Areas

Never try to build a road section over an area that's rolling under the weight of your equipment. Invariably, unsuitable material causes the finished surface to break up under the load of traffic. Occasionally you'll discover small soft areas after the aggregate base has been trimmed and the road or parking area is ready to pave. Use a plug of asphalt up to 6 inches deep to bridge these small soft areas. Asphalt will bridge even a very soft area if it's thick enough. But remember that the asphalt plug must be placed well ahead of the paving equipment so that the plug is hard by the time the top surface is put down. Place the plug the day before paving whenever possible. And keep heavy loads from traveling over them until at least the day after they've been paved over. A few small unsuitable areas may appear when you're trimming subgrade. If they seem to be shallow, the grader operator can roll the unsuitable material across the subgrade far enough to mix it with dry dirt. After it's mixed, blade and roll it back into the same area.

Working in Bridged Areas

Occasionally you'll find an area of unsuitable material under a layer of firm earth. This earth usually bridges the unsuitable material except for a few isolated areas where the equipment has broken through. This situation can be troublesome for any contractor. Usually, loaded scrapers pulling out of the unsuitable area will break off more of the bridging layer. The unsuitable area will continue to get larger at each end of the run as the scrapers enter and leave the area, exposing more of the formerly bridged area on each end with every trip they make. Resolve this situation in one of two ways.

For sizeable bridged areas— If most of the unsuitable soil can be removed without breaking through too much of the bridged area and doubling the size of the unsuitable area, use scrapers or a loader. Remove all unsuitable material except for where the bridged area on the ends has collapsed from the equipment weight.

Then let a hoe take over. The hoe removes the remainder of the unsuitable material without additional deterioration of the bridged area from heavier equipment.

For thin bridged areas— If the bridged area is thin, you can't use scrapers to remove isolated unsuitable areas. The weight of the scrapers would cause the thin bridging to collapse, resulting in larger unsuitable areas. In this case, use a hoe to remove the isolated unsuitable areas. The hoe should load the unsuitable material into small dump trucks with a weight capacity less than the scrapers. The unsuitable material can be removed without breaking through the thin bridging areas. This reduces the chance of increasing the size of the unsuitable areas before base can be placed.

The hoe methods work well in parking lots where only car traffic will travel over the finished asphalt and not truck traffic. Sometimes the ground is firm enough to hold a track loader even though scrapers are breaking through in the bridged area. If this is the case, use a track loader in place of a hoe to increase production. This method of removing unsuitable material should only be tried if the surface bridging the unsuitable material is thick enough to hold up during the rocking and paving operation.

Filling After Removing Unsuitable Soil

When filling the area after removing the unsuitable soil, be sure to use equipment that's light enough so that it doesn't break into more of the bridged area around the edges. A small dozer or backhoe with a front loader bucket is ideal for this work.

The inspector may direct the contractor to incorporate the unsuitable material removed into the fill area. If so, the scraper operators must dump only thin layers of soil. Spread the unsuitable material in a thin, even layer and then add a layer of good material. Good material draws moisture from the unsuitable material when the two are mixed. No water truck is needed. The moisture from the unsuitable material should provide plenty of moisture.

Sometimes the unsuitable material is so wet that you disrupt the filling operation by trying to incorporate it in the fill. If so, you should be compensated for the production loss. If there is more unsuitable material being directed to the fill than can be incorporated without causing the fill to become unsuitable, point this out to the authority involved. An inspector or authority who insists on dumping unsuitable material into a fill does so at his own risk. If the entire fill becomes unsuitable because of the inspector's error, you're not to blame.

During hot and dry weather, shallow areas that are pumping slightly should dry out enough in one day, especially if they've been ripped. Work them the following day.

Earth that is almost unsuitable will form slight grooves or ruts as the trucks haul aggregate or pavement over it. Roll these ruts flat before spreading aggregate or pavement.

Unsuitable Soil Around Utility Lines

Often you'll find unsuitable material around water, gas, sewer, telephone, electrical, or drain lines. This is common when replacing old pavement with new. The asphalt or concrete of the existing road tends to trap water and saturate the soil underneath. If there are utility lines under the existing pavement and the ground below is soft, there's a real danger that the heavy equipment will break the lines. A loaded scraper can compress the soil enough to break a water main 5 feet deep.

This is probably one of the most aggravating excavation problems. In a short time, water floods the area to be excavated and no work can be done for hours or days. You can avoid problems like this by locating all the utility lines before starting the job. Know both where they are and how deep they are. This is doubly important when unsuitable material is involved.

Use the Right Equipment

The best piece of equipment for removing unsuitable material above and around utility lines is a backhoe. Sometimes you can use a small dozer to push the unsuitable material out to where scrapers can pick it up. The equipment you'll choose will depend on what utilities are involved and how much weight they can take without damage. A reinforced drain line can take much more abuse than a vitrified clay sewer line or asbestos water main. A cast iron water main is a sturdy pipe but will snap in the center with excessive weight or a sudden jar. Telephone ducts are as fragile as a clay sewer line. A gas main can probably take more weight without damage than any other utility line.

If the unsuitable material must be removed below the utility lines or very close to the top, a man on the ground with a shovel and prod rod should direct the hoe operator. Use the prod ahead of the work to keep the hoe operator from damaging any service lines attached to the main line. Water main services usually come from the top half of the main and you can easily hit them if you're not careful.

Notify the Utility Companies

Notify the utility companies before starting to excavate. They are very helpful in locating the lines and service connections. Some utility companies do the locating for you while others will supply you with a utility plan of the construction area. Remember, most utility companies require 24 to 48 hours notice before construction begins.

Some states have an underground alert number to call that informs all the utility companies that belong to it that construction excavation work is beginning. If there is an Underground Service Alert (U.S.A.) number in your area, call it, but be aware that not all utility companies belong to the service.

Removing unsuitable material around utility lines is a slow process. Don't rush it. It's an expensive way to excavate. Still, it's much cheaper to take the time to work around the utilities carefully rather than damage them. The cost of repairing a telephone cable or gas main is high. Cutting a gas main or an electrical line is both expensive and dangerous. Carelessness can kill someone.

Backfilling Around Utilities

Most utility companies demand that you place 1 foot of sand backfill over their lines before backfilling with excavation material. Where the line is extremely shallow, the utility company may decide to lower the line or pour a concrete cap over it for protection. You should be reimbursed for any sand used. And the contractor shouldn't be charged for any lowering of the lines unless the job specifications require it as part of the job.

After all the unsuitable material has been removed, the inspector usually decides on the kind of fill material to use to bring the road back up to the subgrade level. Generally, it'll be dirt, cobbles, pit run gravel, or aggregate subbase material. After the unsuitable material has been replaced, trim the area to finished subgrade.

Construction Fabrics

Woven and non-woven synthetic fabrics such as polypropylene and polyester are now used on many excavation jobs. They're commonly called *filter fabrics* and are very strong and relatively cheap. These fabrics are used for three purposes: to provide stability, to separate different materials, and to filter solids out of water passing through the fabric.

When a subbase is so saturated that it's a quagmire, only a dozer with mud tracks or a hoe can be used to level the roadbed. And, of course, saturated soil can't be compacted. Fabric may be the answer. Covering the soil with fabric may give enough stability to permit work to continue. Laying fabric can save time and money and may be the only cost-effective solution available.

Using Fabric on Unsuitable Areas

If the subgrade is soft, fabric does no good if it's directly under the roadbed. It has to be covered with enough fill to provide a stable base. Fabric keeps the material you place over the unsuitable area from sinking down into the soft grade. It also keeps the soft grade below from working up into the material you put over it. If the grade is a quagmire, it may take 3 feet or more of dirt or base material over the fabric to stabilize the area. But laying 3 feet of dirt or base will probably be cheaper than removing 4 or 5 feet of unsuitable material and filling with cobbles or aggregate to stabilize the area.

If you notice a slight pumping in the subgrade, fabric and 8 inches of base may be enough to stabilize the area. The design engineer or soils engineer usually decides whether to use fabric and how much base material is needed to stabilize the area. In some cases you may have to both remove the unsuitable material *and* lay fabric.

If there's a soft area that is 5 feet deep from subgrade down, the engineer may have you do the following: Excavate the area to 2 feet below subgrade and lay construction fabric over the remaining 3 feet of unsuitable material. Laying 2 feet of fill over the fabric and then laying the road section aggregate will usually stabilize the soil. This eliminates the extra work and expense of excavating the remaining 3 feet of unsuitable material. The savings in excavation time will more than pay for the fabric.

Placing Construction Fabric

When placing construction fabric, bury the fabric under a layer of fill before running heavy equipment over the area. Otherwise heavy equipment can suck fabric into the tracks or wheels.

Look at Figure 17-1. Here you see a pile of rock that has been left for later use. It'll be pushed ahead when the next section of fabric is put in. Notice the perforated plastic pipe in the center of the road section. The pipe will drain away any accumulated water. Filter fabric is used over and under the pipe to keep soil from plugging the holes.

Handling the Fabric

Overlap the edges of the fabric 1.5 feet or more when several widths are needed. Stake the fabric down or throw gravel along the edges so wind doesn't blow it around. The engineer may require

Filter fabric and pipe
Figure 17-1

that the seams be sewn by a hand-held sewing machine rather than overlapped.

Two men can usually handle the rolls unless the fabric is wet. When wet, they're very heavy. Keep them in a dry area or covered.

You can cut the fabric with a sharp knife, but it quickly dulls the blade. Keep a file or sharpening stone available.

Fabric Isn't a Water Barrier

Remember, construction fabric isn't a water barrier. Standing water will pass through the fabric and saturate the upper base material. You still have to keep the water out. If the engineer feels that water may be trapped under the area later because a soft condition was found during excavation, he may design an under-drain to remove possible seepage. Look back at Figure 17-1. A 4-foot wide strip of fabric was laid down the center of the road in a slight swale. Plastic perforated pipe was laid over the fabric. Then, the road base fabric and aggregate were laid over the pipe.

Uses for Construction Fabric

Erosion Control

Construction fabric is excellent for erosion control on channel and river banks. Use it as a barrier between soil and shot rock installed as rip rap. The fabric allows water to pass through but not soil. Fabric is resistant to mildew or rotting.

Construction fabric works well on a cut slope where erosion control is needed. Spread it on the slope and stake it down. Then spray asphalt emulsions on the fabric. While the fabric is still wet, sprinkle seed on the oil. The emulsions will hold seed in place until it sprouts. The roots of the grass will grow through the fabric, binding it to the slope. Hydroseeding over the fabric also works well.

Under Asphalt Overlays

Construction fabric is excellent under asphalt overlays on runways, roads, and parking areas. It helps eliminate reflective cracking and stress cracking that cause water penetration. Eliminating cracking should add years to the life of the asphalt overlay.

When putting fabric under an overlay, clean and oil the existing pavement before the fabric goes down. With a tractor especially equipped to hold the fabric roll, lay the fabric over the oil. Be sure this is done as soon as the oil is spread and before any traffic can run on it. Using a tractor with roller grips to spread the fabric helps keep it smooth.

In Figure 17-2 you see a tractor rolling fabric over oiled pavement. Sometimes a rubber tired roller is used to press the fabric into the oil. Asphalt trucks dumping asphalt and the paving machine should be close behind. The streaks you see in the pavement in Figure 17-2 were made by the profiler which ground off 2 inches of the existing pavement. Notice the 2-inch lip at the curb to the far right. Next, 2 inches of new pavement will be laid.

Tractor rolling fabric over oiled pavement
Figure 17-2

Efficient profiler machines can now mill existing asphalt or concrete with relative ease. This has eliminated the need for demolishing and rebuilding old roadbeds. It's more cost effective to mill the old surface down to the desired elevation, then oil the surface, place fabric and repave. The old milled asphalt is recycled at the asphalt plant and reused.

Subsurface Drains
Construction fabric is ideal for subsurface drains because it's highly permeable to water but keeps dirt from passing through. Put the fabric in the trench and lay pipe on it. Place permeable rock material over the pipe, leaving enough fabric to cover the top of the permeable rock so no dirt filters down.

Filter fabric under railroad ballast
Figure 17-3

Under Railroad Ballast
Construction fabric is now widely used under ballast for railroad tracks to keep the ballast in place. You can see how it's done in Figure 17-3. This reduces movement of the soil so less ballast is needed.

Miscellaneous Uses for Fabric
Other uses include: 1) Loose casing around pipe to be encased with concrete. 2) Against underground building walls to protect water sealant. 3) To wrap pipe to protect coatings during backfilling. 4) For blasting mats. 5) Under the surface of horse arenas to keep dirt and mud from penetrating the riding surface. 6) Erosion fence. Always keep these fabrics in mind when working unsuitable soil or when erosion is a problem. Construction fabric comes in several weights, widths, and lengths. Your local supplier will be happy to supply more information.

Compaction

This chapter describes good compaction practice and shows how compaction tests are made. I'll also highlight some problems that are common when your work requires compaction testing.

There are many types of compaction testing equipment that give accurate readings. There are also many opinions on the type of equipment that gives the best compaction. One superintendent might prefer a dozer and sheepsfoot roller on his dirt spread job. Another wants only a self-propelled pad drum roller. Most contractors agree, however, that good compaction is the result of controlling the amount of compaction effort and water on each layer before it's covered with another layer of soil.

Learn from Experience

Good compaction requires more experience than any other type of excavation work. It's difficult because there are so many types of soil, each needing a different compaction technique. For example, a sandy soil needs much more water than a heavy clay before it reaches maximum density. You need to know what the different soil types need for proper compaction.

After a few years of experience in compacting different types of soil, you'll be able to look at a particular soil and know whether it has enough water to compact well. One quick test is to grab a handful of soil and squeeze it. Soil that crumbles when you open your hand is too dry. If it holds solid, it should be good. If you can squeeze moisture out of the soil or if it feels sticky, it's too wet.

Compaction Testing

The amount of compaction is a measure of the density of a soil. The more dense the soil, the greater the load it will support. Most roads and buildings are designed on the assumption that the soil has a certain density or load-bearing capacity. The job specifications usually require the excavation contractor to compact the soil to the density specified by the designers. Soil tests will confirm if the soil has been compacted adequately and whether it'll support the planned road or building.

The two most common types of compaction tests are the *sand cone* and *nuclear* tests. The sand cone test is the oldest method of testing. Most testing firms say the sand cone method is the more accurate of the two. However, the nuclear test method has generally replaced the sand cone test method. It's now more common and much faster as well.

Sand Cone Test
To make a sand cone test, dig a round hole with a volume of 1/10 of a cubic foot. Weigh the dirt from the hole. Then pour from a cylinder of sand with a known weight into the hole until the hole is full. Weigh the remaining sand to determine the amount of sand in the hole. Now the precise volume of the hole is known. Seal the soil removed from the hole and take it to a soil testing laboratory. There it'll be dried and weighed again to determine how much of

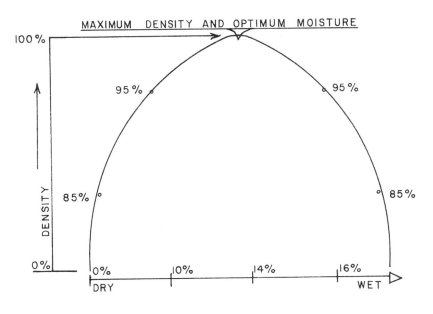

Moisture density curve
Figure 18-1

its weight was water. Take an additional sample of soil from the same spot to the lab so they can plot a moisture density curve on the soil.

Plotting a moisture density curve— The lab will add known amounts of water to samples of dry soil. Then they'll place the soil in a tube, tamp it a given number of times, and weigh it. Adding water to the dry soil and compacting it increases the density until it reaches optimum density at a certain percentage of moisture. The best moisture level is the volume of water the soil sample contains at the point when it reaches maximum density. Each time a sample is pounded under a given moisture content, a certain soil density will result. The percentage of moisture and density is plotted as a point on a graph as shown in Figure 18-1. After several samples are plotted, they're connected with a curving line. That's the moisture density curve.

The highest point on the curve is called the point of *maximum density and optimum moisture.* That point is considered 100 percent, or optimum compaction. You may see it called 100% A.A.S.H.O. If the lab sample maximum weight is 150 pounds at maximum density and the field sample weight was 145 pounds, then the test result would show 96 percent compaction (150 pounds divided into the 145 pounds field weight and rounded down to the next whole number).

Nuclear Testing

There are two ways to take a nuclear test. With one method you simply set the box on a smooth surface and read the instrument. The second type requires that you drill a small 1/2-inch hole slightly deeper than the section being tested. Then stick a test rod attached to the box into the hole. Each device works the same way. Nuclear impulses are sent into the soil or aggregate base. A gauge on the box records the impulses reflected by the soil and returned to the machine. The better the soil is compacted, the fewer impulses are received and the lower the reading on the gauge. In tightly compressed soil, fewer nuclear pulses return to be counted.

Figure 18-2 shows a nuclear density tester. In less than five minutes, this tester will display the numbers the technician needs to compute density and moisture content. The rod extends through the gauge into the ground.

For example, if the soil is compacted to 95 percent, the reading might be 22,000. If the soil is compacted to 90 percent, you might get a reading of 26,000. Lab personnel know the density of the material from the reading. They have a chart that gives the particular characteristics of that nuclear tester. This chart is based on the readings the tester gives when placed on a block of material of known density, like concrete. The soils tester in the field usually carries a polyurethane block to check his gauge in the field. The nuclear tester should be checked occasionally to be sure it's operating properly.

Nuclear density tester
Figure 18-2

The soil technician gets a sample of soil at each nuclear test location, takes them to the lab and mixes them together. From this soil, you can get a moisture-density curve using the same method as the sand cone test curve. From this curve, the lab establishes the weight of that soil at its maximum density. Once the maximum density weight is known, the lab determines what the nuclear gauge reading should be at the density required by the specifications.

The Importance of Water

Water acts as a lubricant and helps the particles of soil slide into place. If too much water is added, the particles of soil tend to float, lowering the soil density. On the other hand, if the soil is too dry, the particles won't slide into the small voids and the density will be lower.

If the soil doesn't contain the correct amount of water, it won't pass compaction testing no matter how much you roll it. If soil fails the test because it's too wet, it can be rolled again after it dries some. It'll probably pass then. If the soil didn't pass because it was too dry, you'll just have to rip it up again, add more water, and reroll it.

Meeting Compaction Standards

Most embankment fills must be compacted to 90 percent. Generally 90 percent isn't difficult to get. On a large job, a greater depth of soil may be put down at one time than the compactor can compact. Still, the soil may pass the density test if the spread was done correctly. Use the scrapers hauling in the fill to run over the fill they just dumped on the previous pass. That way the scrapers compact the fill as they haul dirt. Even the water truck can do some compacting. Anything running on the fill adds to the compaction. That's why it's much easier to get good compaction in a large fill area than in a small confined area.

For example, assume an area 200 feet long and 30 feet wide receives 3,500 cubic yards of compacted fill during one shift. An area 800 feet long and 300 feet wide could easily receive 7,500 cubic yards of fill using the same compactor but using the scrapers as compactors.

POORLY   GRADED

WELL   GRADED

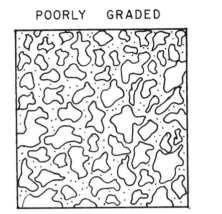
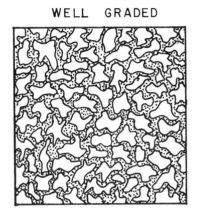

Material gradation
Figure 18-3

Some expansive clay soils may require just 86 percent compaction, using only a disc and dozer. This is usually when a high optimum moisture content is required.

Meeting Subgrade Compaction Standards

You'll have more trouble with density tests during subgrade preparation because the standard is usually 95 percent density. It's much more difficult to get 95 percent than 90 percent. To get 95 percent, the soil must be mixed very well. If there are several types of soil in the same fill, as is usually the case, it takes more working to get the proper compaction.

For example, the same fill might have hardpan chips, clay, and sand combined. A few chips of hardpan or a small sand pocket at the point tested can cause the soil to fail. If you're working in a combination of these materials, mix them thoroughly. All the chips of hardpan and sand layers must be mixed with the clay, and the hardpan chips must be crushed. Figure 18-3 shows how a well-graded soil mixture will look.

Clay and sand together make a bad combination for compacting because clay requires only a small amount of water, and sand requires a great deal of water. It's difficult to get the two mixed together without getting too much water in the clay (causing it to pump) or not enough water in the sand.

To get 95 percent compaction, the water content must be just right. If you notice any movement in the soil while compacting, you added too much water. Going over optimum moisture by a few percent may still pass a 90 percent compaction test. But with a narrower leeway, you won't pass a 95 percent test until the excess moisture dries out.

What does it take to consistently pass compaction tests at 95 percent? A lot of experience and trial and error experimentation. Supervise carefully any time your crews are compacting soil. That's the only way to gain experience.

Compacting Aggregates

A vibratory roller is excellent for compacting nearly any aggregate in up to 6-inch layers. In most cases, aggregate base material or road rock is easier to compact. But you may have a problem with aggregate material that doesn't have enough sand or rock dust in it. Aggregate that contains less than 6 percent of particles passing a number 200 sieve is considered very "clean." It probably has too much washed rock and not enough crushed rock to bind well. When this is the case, saturate the aggregate with water and roll it vigorously to reach 95 percent compaction.

With some soils or aggregates, the subgrade or base will be so soft after achieving 95 percent that you'll have to leave it idle to dry out before it can be paved. This may not seem logical, but it may be perfectly acceptable from a soils engineering standpoint.

When you're trying to compact soil or aggregate material to 95 percent, avoid putting down more than 6 inches on any one lift. Sometimes, you'll have to compact layers only 3 inches deep to achieve 95 percent.

Use a sheepsfoot, grid or pad drum type self-propelled roller to roll the subgrade first on any large job. This ensures that there are no dry pockets and that the hardpan chips have been smashed to a size that allows easy mixing. Then use a multi-tired pneumatic or vibratory smooth steel drum roller. Figure 18-4 shows a pneumatic

Pneumatic tired roller
Figure 18-4

tired roller used as the required second roller for asphalt concrete paving. This machine is excellent for rolling trimmed subgrade with heavy clay content.

Compacting Clay Soils

A rubber tired roller works well in clay soils. Clay won't stick to the rubber roller wheels as it may to a steel drum roller. But a vibratory steel roller will get compaction much faster, so it's often the preferred choice. For rolling aggregate base material, a smooth steel drum vibratory roller is the fastest. If the job is large enough for two rollers, a smooth steel drum vibratory roller followed by a multi-tired pneumatic roller is an excellent combination.

You can add water to dry material and work it in with a pad drum roller. If you use a multi-tired pneumatic roller or smooth drum vibratory steel roller, mix the necessary moisture in the soil before rolling because these rollers seal the surface. Very little water penetrates after the first pass over the soil.

Selecting the Right Equipment

Of course, the job size determines the type of compaction equipment you'll use. Selecting the right equipment is very important. If you're working on a large fill job where soil gradation is a problem because there are several types of soil, consider using a disc to mix the soil fully. You can use a dozer to pull a sheepsfoot roller and a disc. What the sheepsfoot misses, the disc will be sure to mix thoroughly. You can also attach a disc attachment to some self-propelled compactors. Working with a disc on a self-propelled compactor is easier if you use a hydraulic ram to raise and lower the disc.

On a small excavation and fill job, it isn't practical to bring in equipment to mix the soil. You can generally get by with only a grader and smooth drum type vibrating roller. The grader can do some mixing as it levels the soil so it can be rolled. This way you won't have to change equipment when you're ready to compact the subgrade and aggregate base. The same two pieces of equipment can be used for both compaction of the existing soil and building up the road cross section.

Compacting Promptly Is Important

Once you've excavated a parking lot or road to the subgrade level, and as long as it has enough moisture, compact the soil immediately if possible. When you return to compact and trim the subgrade, chances are you'll find that it has dried out. You'll have to rip it up and water it before you can start compacting.

If there are no underground utilities to be placed, consider compacting the initial excavation while the natural ground moisture is about right for compacting. A vibratory pad drum roller would be excellent for getting 95 percent compaction in a case like this. You can leave the area for some time after compaction with little loss of moisture because dense soil loses moisture very slowly. Even though the surface will look dry several days later, usually neither the compaction nor much moisture will have been lost. The surface can be watered, trimmed, and rolled with very little additional compacting.

Use this same principle when working fill areas. If the equipment is available, never allow a fill to dry out if you have to return later to compact the top 6 inches to 95 percent. Confer with the soils engineer. If the grade is close, he may accept it as it is.

Work in Several Areas at a Time

When making several small fills, it's easier to work in several areas at a time. This way the scrapers, water truck, and compactor have more room to operate and they'll be more productive.

Here's how to organize the work: Rip, water and compact the original ground in the first fill area. While the scrapers are dumping a layer of fill across the first fill area, rip, water and compact the second fill area. Once the scrapers have spread a layer of fill across the first fill area, have them start dumping in the second fill area. Then have the compactor and water truck move to the first fill area to water and compact the layer of fill that was spread by the scrapers.

When the scrapers have spread a layer of fill across the second fill area, they again haul to the first area which you've just compacted. As the scrapers move back to the first fill area, the compactor and water truck move to the second fill area to water and compact it. This procedure continues until each fill has been brought to grade. When you work two or more fills this way, all equipment can work at its maximum rate with no waiting.

You'll need to rip some soils before compacting them. Ripping is needed to loosen the soil so you can mix water into it. When compacting subgrade, if moisture is already present in the soil, ripping usually isn't necessary. Rolling should get the compaction needed.

Working in Hardpan

One exception is hardpan. Even though it's very hard, you'll usually have to rip it and recompact to get a compaction density of 90 percent or more. If you've made a good compactive effort at the right moisture level but the tests still don't pass, check with the soils engineer taking the tests. He may not be using the current soil curve. I'll explain what that means.

The soil engineering firm usually computes the soil compaction curve from a sample taken at the beginning of the job. But as the job progresses, the soil type may change — even though the soil looks the same. If the soil firm doesn't take another sample, the compaction test results could be wrong. If tests begin to fail even though you're following the same procedure, ask the soil firm to take a new sample.

It's possible to over-roll some soils and aggregates. After three or four passes with a vibrating roller, the area being rolled may feel firm and look tight. More rolling may cause separation of the top layer. If this starts to happen, turn off the vibrator and give the soil a shot of water. After the water has set, flat roll with the vibrator off. This should restore the firmness. Lack of moisture in the top layer will also cause the top layer to ravel and look over-rolled.

If you suspect over-rolling, dig up a small section of the top layer and check the moisture content. If the soil seems moist, the problem may be too much rolling. If the soil is dry, add water and roll again.

If most of the tests are good but a few are low, even though the moisture's right, try rolling once more at a right angle to the first passes. With vibrating steel drum rollers, hard sections of soil can create a bridge, sheltering some areas from full compaction. The entire area under the drum may not get full roller pressure. Cross rolling should eliminate this problem.

Trench Compaction

Subsidence over buried utility lines is a common problem. Fortunately, a little care and knowledge about compacting will prevent most subsidence in trenches. In many soils, all that's needed to compact trenches is water jetting. But water jetting won't be effective if backfill was placed incorrectly.

Trench Jetting
Here's the first rule: If a trench will be jetted, never fill it to the top or heaped over the top with dirt. Look at Figure 19-1. It shows the correct and incorrect ways to backfill a trench before jetting. The unfilled lip in Section A serves four purposes:

1) It helps the person jetting to see the edge of trench.

2) It helps the trench settle faster because of its loose nature.

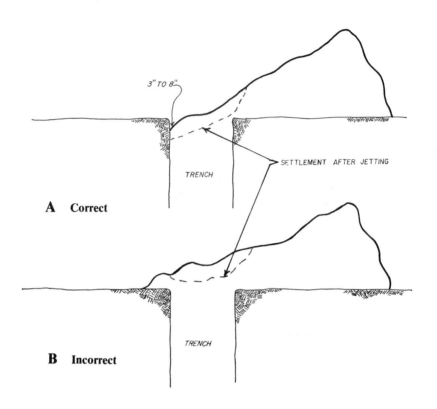

Backfilling a trench to be jetted
Figure 19-1

3) If excess water flows from the jet rod to the ground surface, it will be trapped in the trench. Otherwise it can flood the area where the person jetting must stand or walk.

4) After the trench settles, it will leave room for a dirt cap of moist dirt to be pushed in and rolled over the jetted trench.

The Jet Rod
The length of the jet rod is important. It should be long enough to reach near the bottom of the trench, but not so long that it will hit

and damage the pipe. The distance from the bottom will vary with the trench depth. For a trench that's 20 feet deep, use a 15-foot jet rod. That's long enough because the weight of the fill above the rod end will force the water down to the bottom of the trench.

As the trench becomes more shallow, the pipe should reach closer to the bottom. For instance, on a 10-foot deep trench, the jet rod should be 7 feet long. If you use a 5-foot jet rod, the water may build up enough pressure to come to the surface because the fill above doesn't have enough weight.

Jetting Techniques

Always start jetting at the downstream end of a trench and jet upstream. First, plunge the jet rod the full depth. As you notice settlement, pull it up to intermediate levels until the jet rod is within 2 or 3 feet of the surface. Don't remove the rod until trench settlement is obvious to the eye. Usually at that point you can see standing water about 2 feet down.

Some water can be left to run on top if the dirt is very dry or sandy. This method must be repeated, moving the jet rod ahead upstream each time. The distance you'll move the jet rod ahead depends on the width and depth of the trench. If the trench is 4 feet wide and 12 feet deep, stagger the jet rod placement from side to side and move it upstream 5 or 7 feet each time. On the other hand, in a trench 2 feet wide and 6 feet deep, the rod can be placed in the center of the trench and moved upstream 6 or 8 feet each time.

The distance the jet rod should be raised up and moved ahead will vary with the type of soil being jetted. Only experience will eventually make the choice easier.

Jetting in Fall and Winter

Jetting in the fall or winter poses more of a problem because the ground won't dry quickly. Lower temperatures and more moisture in the air slows the drying at that time of the year. Many trenches fail because the contractor didn't use the correct amount of water needed for good settlement. And there's a reason for that. If the trench is too wet, it will delay the grading and paving that follows.

Jetting a partially-filled trench
Figure 19-2

The contractor may feel that mechanical compaction is too expensive, so he just does a light jetting. The trench fails later from water seeping into the trench after paving has been completed.

There's a way to avoid this problem without adding the expense of mechanical backfill. Leave the initial backfill down 2 feet, then jet normally. After jetting, backfill the remainder of the trench and wheel roll it with the grader. That creates a firm cap over the jetted trench. There are two requirements for the dirt cap that's wheel rolled: First, don't use lifts greater than 12 inches. Second, make sure it has enough moisture mixed in it for good compaction. If the trench is over 2 feet wide, leave the trench backfill a little lower than 2 feet from the top. And have an experienced man doing the jetting so it isn't over-jetted. A person jetting for the first time should take the time to watch an experienced jet person first. Talk to the jetters about the jetting characteristics of the soil in your area.

Look at Figure 19-2. This trench down a roadway is being jetted after it was backfilled to within 2 feet of the top. The top 2 feet will be filled and wheel rolled so a temporary patch of cold mix can be placed on it. Then traffic can drive over it until the next shift begins.

Mechanical Compaction

There are many types of compaction equipment. The equipment you use will usually depend on the space you have to work in. Working in narrow trenches when 90 percent compaction is required is slow going. For maximum compaction, any material dumped into the trench must be leveled before tamping. When 90 percent is required, you usually shouldn't place more than 6-inch lifts before tamping. In Figure 19-3, a trench is being compacted with a gas-operated foot-type tamper.

In narrow trenches, it's important to mix the moisture needed for compaction into the material before it's dumped into the trench.

Mechanical Compaction Equipment

The type of equipment used for trench compaction depends mostly on the width of trench being compacted. Here are a few of the options for a narrow trench: the Hoe Pac, a foot tamper, a plate tamper, or a remote-controlled double drum roller. On a wide trench, a small dozer pulling a vibratory sheepsfoot roller or a pad drum roller are good choices. See Figures 11-1 and 11-2 in Chapter 11.

To increase production, use larger equipment to place the fill at the fastest rate. Using the dozer and sheepsfoot or pad drum roller with dozer attachment, you can mix some water into the material in the trench. Even with this operation, it's always faster if some water is added to the material as it is pushed into a wide trench. Always keep any area being compacted as smooth and level as possible for maximum compaction.

Protect the Pipe from Compaction Equipment

There's one concern you should always keep in mind when using mechanical compaction. Be sure there's enough fill over the pipe

Compacting a trench with a mechanical tamper
Figure 19-3

to keep it from being damaged by the compaction equipment. Cover the pipe deep enough with gravel, crushed rock, or sand slurry. If necessary, hand tamp before beginning the mechanical compaction.

Consider using aggregate to fill the entire trench for short crossings when working in urban areas where traffic and time are problems. Aggregate compacts much easier and faster. Some agencies will allow crushed rock to be used within 1 foot of the top of trench and require no compaction of the crushed rock. The last foot is then capped and compacted with aggregate and asphalt as required. Ask the inspector or soils engineer what options you have for material use. Then choose the one you feel you can work with best.

Lime Treated Base

When the soil is heavy clay, the job specifications may call for a lime treatment. In this chapter we'll look at the three methods for mixing lime base, and outline the sequence in which the work should be performed.

Trimming the Subgrade

The need for lime treatment doesn't affect the dirt moving operation. The only difference is in the trimming. When a lime treatment is to be applied, grade the subgrade material as precisely as possible to the correct level before adding the lime. That way large amounts of lime treated base won't be removed in the finish trimming.

A little more time spent trimming the subgrade before the lime is added will save time on the final trimming. Usually plus or minus 0.08 foot is the tolerance allowed for lime treated base. It's a good idea to trim the subgrade to the same tolerance. Adding the usual 3 to 5 percent of lime does not add enough volume to affect the grade level significantly. The shrinkage caused by compacting will usually offset the volume of the lime added, unless you're treating over 8 inches of soil. If so, consider undercutting a little.

Spreading the Lime

When the subgrade has been trimmed, the lime spreading operation can begin. The first step is to rip up all of the road surface or parking area to be limed. Then use one of the following three methods to mix the lime.

1)   Grade the ripped-up soil into a windrow, flatten or trench the top, and add the lime over the windrow. Then thoroughly mix the lime and soil in the windrow by rolling the soil across the road with the blade of a grader several times. Add water as it's being mixed. This is a slow method but it's acceptable on a small job.

2)   Mix the lime and soil by running a pug mill over the windrow. The grader then spreads the mixture as water is added. A *pug mill* is a rectangular box on wheels with two rows of power driven steel arms that churn the dirt and lime together as it is pulled along straddling the windrow.

The problem with these first two procedures is that water must be added to the material while mixing and spreading. If all the required water is added before mixing, the material will tend to ball up, making mixing very difficult. Usually the lime can't be mixed properly when the soil is too wet.

3)   The fastest way to mix in lime is to use a machine built for the purpose. This machine is similar to a large rototiller. It chews up the soil as it adds water and mixes the lime and soil together. Look at Figures 20-1 and 20-2. In most cases the mixing can be done in two passes.

Even when a lime mixing machine is used, the soil should be ripped before the mixing begins. Then the lime is dumped on the ground at a set rate ahead of the machine. A water truck must move along at the side or ahead of the mixing machine to supply

The lime is rototilled as water is added
Figure 20-1

The front engine drives the water pump on this recycler
Figure 20-2

water for mixing. Most lime mixing machines have a gauge which indicates to the operator the rate at which the water is added. In most soils, 15 percent water must be added to start the action of the lime.

In all three operations above, you must have some method of measuring the exact amount of lime being dumped. There are trucks built specifically for spreading lime. They can be set to spread the amount of lime needed. These trucks are covered so the powdered lime won't blow out on the road while traveling from the plant to the job.

Before you begin mixing the lime, anything covered with dirt, such as a water valve or manhole, must be uncovered if it is shallow enough to be ripped into or hit from the rototilling.

Compacting the Lime Treated Base

Use a sheepsfoot or self-propelled compactor with steel pads for the first compaction pass. At first the soil may seem too wet, but the lime has a tendency to absorb moisture quickly. Then use a steel drum vibratory or rubber tire roller for the final rolling and during the trimming. A rubber tire roller usually works well because the material is less likely to stick to the rubber tires.

The lime acts in the soil fairly slowly. After the lime has been mixed, rolled, and the grade shaped, you may have to leave the surface until the following morning before trimming begins. The surface must be kept damp at all times to avoid cracking.

Fine Trimming and Oiling

Complete the fine trimming in the same way you would trim untreated soil. The specifications may require a maximum thickness of only 6 inches per lift. Once trimmed, the grade should be oiled. Oiling the lime base seals in the moisture, keeps the soil from cracking and allows curing to take place. For best results, allow three days for curing after oiling before equipment or traffic is allowed on the grade.

But don't oil the base if another layer of lime base is to be placed over the existing layer. In that case, the lower layer should be kept damp until the second lift is added. If aggregate is to be placed on the lime base, you may not need to oil it. Just water to keep it damp.

Passing the Compaction Test

Even lime base that has been compacted well and tested soon after trimming may fail the compaction test. If the test results show compaction of 93 percent or better, it's usually safe to go ahead and oil the surface and call for another test after the lime subgrade has had 18 hours to set. Lime treated base often picks up 2.5 percent more on a compaction test after it has another day to cure. Be sure to lay out a few small squares of plywood before oiling so the testing crew will have a few spots without oil where they can take their tests.

Most lime treating specifications call for all mixing to be complete within 72 hours after the initial spreading of lime. Final compaction should be completed within 36 hours of the final mixing.

Don't attempt lime treatment if the temperature drops below 35 degrees F. Lime takes longer to set up hard in cold weather.

Using Cement Instead of Lime

Cement is preferred if the soil has a high sand, rather than clay, content. Cement treated base may be specified mixed with aggregate, also. If you're using cement instead of lime, follow the same procedure with two exceptions. First, you don't need as much water. Second, finish the fine trimming and complete the final rolling the same day. Usually the final compaction must be completed within two hours of the mixing. Cement acts much faster than lime.

Cement treated base must also be oiled to avoid cracking. If other layers are to be added, keep the surface damp by watering to avoid cracking if no oil is used.

Aggregate Base

In this chapter I'll emphasize the importance of a controlled dumping for aggregate base. I'll also list the steps that must be taken to get a good finished road base. There are two factors that affect the quality of the job. First, the speed at which aggregate base is finished: The fewer times it is bladed, the more fines the surface will retain, leaving a smoother, more firm finished surface. Second, the experience of the grader operator and the paddle wheel scraper operator for this operation. An inexperienced operator making one pass with the paddle wheel scraper can ruin the trimmed grade if he can't pick up the final trim windrow without cutting into the finished grade.

Setting the Grades

The surveyors for most agencies on road jobs will either set bank plugs or grade hubs (blue tops) for trimming the aggregate base. Be sure they're notified in advance to set the grades you need.

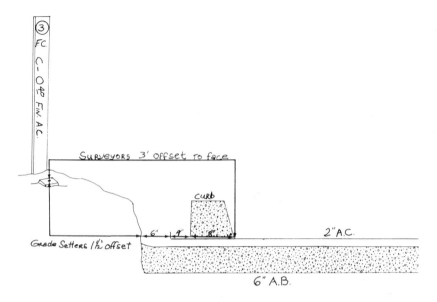

**Surveyor's and grade setter's offsets
Figure 21-1**

When extruded curb will be poured on the asphalt in parking lots, you can usually use the original grades set by the surveyors for grading and placing aggregate base. The grade setter will mark boots and shoot from them with an eye level, or use a laser to set elevations shown on the plans.

The distance to curb marked by the surveyors must be given close attention when cutting grade for extruded curb or barrier curb. The surveyors may mark one of two distances — back of curb or face of curb. The grade setter must determine where he must measure to for his cut. For example, if an extruded curb is called for and face of curb is indicated on the survey stake, he must subtract the width of the curb and some extra distance behind the curb so the curb machine will have a level pad of asphaltic concrete to run on.

Figure 21-1 shows a surveyor's hub that indicates a 3-foot offset to face of curb. Notice the grade setter excavated 1½ feet behind the face of curb. This is to allow room behind the curb for excess aggregate and asphalt so the curb can be poured on a firm base.

The Trimming Crew

A trimming crew should consist of a foreman, a grade setter, a laborer for hopping guineas, a laborer to clean curbs and edges, and operators for a water truck, paddle wheel scraper, roller, and grader. If two graders are doing the trimming, one scraper should be able to handle the job, but one more roller and one more guinea hopper may be needed.

Placing the Base Material

Laying rock base of any kind should be a relatively easy job. The subgrade has been finished. This gives a finished level to work from. The key to efficient placing of base material is to have a *controlled dumping procedure* set up. For example, most bottom dump trucks average 25 tons per load and can set the gates to spread the load 100 feet. If 100 tons are needed every 100 feet and each truck dumps 25 tons, have each truck dump a row 100 feet long. Four parallel 100-foot long rows spread evenly across the road will leave the right amount of aggregate.

It's important to have even, continuous rows. This serves two purposes. It insures an even quantity of rock throughout, and it's much faster for the grader operator spreading the rock. In many cases, and especially on large road jobs, a self-propelled spreader may be used rather than a grader. In any case, a good controlled dump is necessary.

In small parking lots, use only end dumps if bottom dump trucks can't maneuver the turns without breaking the curb.

You can use any dumping pattern. It may be best to dump one continuous row from the beginning of the road to the end before beginning any of the next three rows. You could also dump four rows side by side in the first 100 feet.

Spreading the Material

If you're using a self-propelled spreader that runs automatically from a string line or piano wire, no grades need to be set. The line is enough. It's important to supply the right amount of aggregate. You don't want a self-propelled spreader to run short of material. But you also don't want it to be overloaded.

On a small road job where a grader is used, you'll need to set crows feet. *Crows feet* are lath 12 to 24 inches high with a line marked at the height of the rock grade desired. These crows feet will give the grader operator something to go by for both width and elevation while spreading the base. When there are curbs on both sides, a row of crows feet down the center may be desired.

In a parking lot where a barrier curb surrounds the lot and there are planters throughout, the rock elevation should be marked on the curb every 10 to 20 feet so the grade operator and laborer working the edges have a grade to go by.

As the rock base is being spread, it's important to get the required water content in the base before it's sealed by the roller. When the rock has a dark grey or brown color and feels damp, it's wet enough.

Setting the Grade Hubs for Fine Trimming

After the base has been watered, spread, and rolled, the grade setter or surveyors will set grade hubs for fine trimming. The grade hubs to which the grader's blade trims are called *guineas* or *blue tops*. In most cases the required tolerance for base is plus or minus 0.05 foot. The grader operator should have a guinea hopper to help while trimming. The guinea hopper cleans the hubs off after the blade passes over them and sets the guard lath back up. In Figure 21-2, the guinea hopper is locating the guinea and setting the guard lath next to it. The guinea hopper must be quick so the grader isn't delayed.

Dealing with the Excess Rock

When putting down base with curbs on each side, the excess must be picked up with a paddle wheel scraper. See Figure 21-3. The scraper moves ahead slowly while the operator watches the cutting

The guinea hopper locates the guinea
Figure 21-2

Paddle wheel scraper picks up excess base
Figure 21-3

edge and slobber bits closely. He doesn't want to cut into the trimmed rock grade as he picks up the windrow.

On most roads without curbs, you can leave the excess rock on the shoulders to be used there after the paving is completed. If there's too much, though, it must be picked up with a paddle wheel scraper. Try to leave a small trim to make throughout the road length except for a low area at the end. Place the excess at the end that was left low if it can't be left on the shoulders. A slight trim is good because it's faster to trim an area than to stop and make small isolated fills. The trim also removes the loose rock ravelled by the trucks hauling aggregate over base that was spread and rolled while dumping.

Move the Rock as Little as Possible

Another important point while working rock is to move or blade it as little as possible. Whenever you move rock, the fine materials tend to separate from the larger rock. That leaves the larger material without the binder needed to hold the rock in place. If this happens, and if it's noticeable, the inspector will require that you rework the rock to get the fines mixed in again.

Rocks that come to the top with no fines to bind them are often called *bones*. The base would be called a *bony grade*. Water helps fines stay with the rocks while the material is being worked. Water will also help produce a better and tighter grade. The more moist the rock can be kept without making it mushy, the easier it will be to work. The right amount of water is also necessary for compaction.

Compacting the Aggregate Base

On the final trimming, the roller operator might need to wait until the rock dries a bit, especially if you're using a steel roller. Saturated rock will stick to the roller drum, leaving a rough surface. If you don't know whether the base is dry enough to roll, make a short pass with the roller. If it is picking up the rock, stop and wait for the surface to dry more. A rubber tire roller will not pick up the wet rock. A drum-type vibratory roller is fast and is ex-

Small tractor grades in tight areas
Figure 21-4

cellent for compacting aggregate base. If the drum picks up aggregate, a section of cyclone fencing can be attached over the top half of the drum. This will clean the drum as it rolls. Look back to Figure 11-4 in Chapter 11.

After enough vibrating has been done to compact the aggregate, make the last few passes with the vibrator off. That sometimes helps to set up the top 1 inch.

Finish Grading and Oiling

Use your most experienced operators for finish grade work because of the close tolerances that are necessary. When working aggregate in parking lots where there are planters to contend with, use a small grading tractor to work the corners and small stalls where the grader can't maneuver easily. See Figure 21-4. The tractor is grading base rock in a small stall as the grader grades the larger areas. This is also a good method for asphalt paving.

After spreading the base rock in a parking area that has barrier curbs, a laborer should level the rock built up along the edges with a shovel. Then he should tamp along the curbs with a plate tamper to compact the edges. When the blade and grading tractor have made the final trim, the laborer should fine trim along the curb edges and run the plate tamper along the edges again so they'll match the area rolled by the large vibratory roller.

Remember, if the aggregate base has less than 6 percent of particles passing a 200 sieve, you'll probably have to add much more water than usual to the aggregate for it to pass the compaction test. Keep the aggregate damp before and during the paving operation.

Some county and state specifications require that the base be oiled if no other base is to be placed on it. If oil is required, it should be done as soon after trimming as possible so little moisture will be lost from the rock base. Never oil dry aggregate. If the aggregate has dried out, have a water truck make a pass, spraying the aggregate before oiling. This will help draw oil into the aggregate. Before shooting oil on any aggregate base, check with the inspector to see how much oil should be sprayed, regardless of what the specifications call for. Coverage between 0.15 and 0.20 gallon per square yard of surface is the most common.

The Final Steps

Keep this in mind: Asphalt placed on a poorly graded rock base will conform to the rock grade. There will be low areas where water will pond and not reach the drains. Be sure to string line or swede between curbs and guineas to detect any low area or high spots that must be trimmed.

In parking lots with many corners, check them all to be sure there's good drainage, not a hump that will trap water in the corner. Any swales to drains must be rolled from each side to the swale point and not centered with the roller. This would ruin the swale shape.

In parking lots when extruded curb will be placed on the asphalt, the edge of pavement must be marked with paint around the perimeter and all islands. This will benefit three operations:

1)   If weed killer is called for, it will delineate the islands that will be landscaped. They should get no weed killer.

2)   If oil is specified, you've established a line to oil to.

3)   The paving crew will have a line to pave. It will insure they pave far enough out so the curb will have the excess width needed to sit on.

Calculating the Quantity of Aggregates and Asphalt

To calculate the quantity of most aggregates and asphalt, multiply the square feet of the area to be covered by the number of inches to be placed, then divide that figure by 160 to find the tonnage needed.

Working in Mud

Working in mud or soft ground that won't support equipment can be aggravating, even for an experienced operator. But experience can help keep an operator out of trouble while working in these conditions. I'll explain the best procedure for operating several types of equipment in soft areas.

Track Dozers

I'll start with the dozer because it's the most difficult to pull out of mud if it gets stuck. This is especially true on small jobs where there's only one dozer on the job. When the operator sees that he is about to enter a wet area, he should go slowly. If he feels the front of the dozer start to settle, he should stop and back out immediately. That settling is the first indication that the ground is too soft — the engine will lug slightly and the front end will start to settle. After backing out, examine how deep the tracks sank into the ground. If they're deep enough so the mud is hitting the bottom of the dozer, it's clear that the ground is too soft.

If there's a soft area that *must* be dozed out to dry or be hauled off, here's how to go about it. Keep the dozer on firm ground. Start from the edge and work forward slowly. Push the mud ahead of the dozer blade and be sure that the ground below is firm. Don't doze ahead to the point of not being able to get back out. Don't try to move too much material on any one pass. Keep the tracks from slipping and digging in.

Dozing Over Partially Stable Earth

Partially stable earth traps many good operators. An operator may drive in and out over relatively firm ground many times, while it slowly gets softer from the weight of the dozer pumping more water to the surface. If this happens, the tracks will sink a little more each time until the dozer finally gets *high centered.* In most cases the operator won't notice what's happening because he assumes he can get in and out easily. He's not concerned until it's too late.

When working on partially stable earth, don't run in the same track each time you enter or leave the area. Move over slightly in one direction or the other so the same tracks aren't pushed deeper each time. In many cases a small dozer with mud tracks will be able to work an area so soft that a large dozer would get buried. The small dozer in Figure 22-1 is working in soft mud 6 feet deep with no danger of getting stuck.

Scrapers

Many times a scraper can't load through a muddy area where a dozer can get good traction. If you have to use scrapers to move mud, here's how to do it. Move into the soft area with the scraper bowl down. Move ahead slowly until the tires start to slip and then stop. Then move the dozer into position to push the scraper.

There are two things to keep in mind when the dozer makes contact with the scraper. First, avoid spinning the scraper wheels. Second, don't try to cut too deeply because this will cause the dozer

A small dozer with mud tracks
Figure 22-1

to spin its track. Then the dozer may get stuck. In either case, the dozer operator won't be able to push on. When the scraper is loaded, apply slightly more power, still being careful not to spin the tires. The dozer should keep pushing until the scraper operator can get enough traction to pull away. In extremely soft areas, you may have to take only half loads until traction improves.

Always watch for soft spots during dumping or loading. If there is a soft area that must eventually be worked through, don't drive right into the center. Drive along the edge first. Move closer to the soft area with each pass. With each pass you can decide whether it's too soft or firm enough to hold the scraper on the next pass. If it is too soft, the compactor may be able to bridge the area by mixing in dry fill.

If you're working over an area where the mud is 3 feet deep or deeper and the scraper loses traction, *stop,* and wait to be pulled or pushed. If you keep trying, chances are the scraper will settle to a point where getting it out is going to be a major job.

There's a possibility you can work the scraper free *if* you can see that the wheels are not settling any deeper in the mud. This may be the case in mud only 1 or 2 feet deep when there is a good bottom under the soft material. In this case, swing the nose of the scraper back and forth, snaking it out. Another way is to set the bowl down, lifting the drive wheels off the ground. Then turn the nose and set the wheels down on a firmer area. Be careful not to make the situation worse. Pulling a well-mired scraper out of mud can be difficult if it's loaded.

Compactors

Take care when trying to bridge a fill across a soft area. To get a 24-ton compactor across a muddy area that must be bridged, assume that you need 2 to 4 feet of stable earth pushed out ahead of the machine. Move ahead slowly, being sure that the ground under the compactor is well compacted.

A "lugging" engine will be the first indication of trouble. You'll think you're losing power. Actually the wheels are starting to sink into the fill. When this happens, back off and push more stable earth over the unstable area, working ahead slowly again until the fill will support the compactor. This same procedure should be continued until the muddy area is bridged. The bridge is complete when a firm pad of dirt is spread completely across the soft area.

Other Equipment

A track rig is the best equipment to use in mud. But if not used wisely, it can get stuck just as any other piece of equipment.

The foreman must decide which is more cost effective, using a dozer or a hoe or dragline. In a very large area of deep soft mud, a hoe or dragline might be the only equipment that can handle the job. This is one of those decisions where experience is the best guide. And the foreman should always have a metal prod rod with him so he can prod the muddy areas and find out how deep the mud is before work begins.

Always Use Caution in Mud

The same basic ideas can be used in operating any piece of equipment around mud. The main point is to use common sense. Move forward cautiously. Never drive right into the center of a soft area. Be very careful if there are soft utility trenches in the area where you're working. If you have to cross a soft trench, approach at 90 degrees. Never run a wheel or track parallel over the trench. This increases the chance of getting stuck in the soft trench — and it may break the pipe laid in a shallow trench. A little caution can save many hours of extra work.

Always try to keep the area as level as possible so you're not driving into the area over rough ground. Be especially careful if you're driving into a muddy area heading downhill. Trying to back out of mud, *up*hill, makes for a real challenge.

When using a dozer, shovel the idlers clean at the end of a shift so they'll roll free at the start of the next shift. If they're jammed with dry mud, it may wear a flat spot on them. These are the top idlers only, the rollers the tracks roll on and are guided by.

A final word of caution: If the equipment is being pulled from the mud with a chain or cable, keep everyone well back. A chain or cable can break and snap back like a rubber band, severely injuring anyone standing close by.

Working Rock

Working in solid rock usually involves blasting, followed by a shovel and truck operation. Sometimes constant blasting is needed. I'm not going to describe that type of work here. This chapter will cover the kind of job where scrapers *can* be used, though there is enough rock in the soil to make the excavating difficult, or where the rock can be ripped with a dozer and will break up enough to be hauled with a scraper.

Rock is probably the most difficult material to excavate. An inexperienced operator can do a lot of damage to the equipment on this type of job. One spin of a tire can cause a blowout, ruining the tire — usually an investment of several thousand dollars.

An experienced operator will work methodically in a rocky area. A good dozer operator will study the way the rocks lie and place the dozer blade exactly where he thinks the best leverage may be gained. He will always try to work from a level area he has made for himself. Trying to doze boulders from a rough and uneven surface is hard on the equipment and usually a waste of time. The

Slope bar attachment on a dozer
Figure 23-1

operator must be thinking ahead at all times. He must pick out voids and then find the correct size rock to fill that void. He can't just excavate and let the boulders fall where they may. This would be a very inefficient way of doing the work.

Cutting Slopes in Rocky Soil

Slopes on a rocky job should be cut with a slope bar on a dozer. The slope bar and dozer are an excellent choice if an existing slope has to be recut. Figure 23-1 shows a slope bar attachment on a dozer. This is very effective for cutting a rocky slope or in wet conditions where a grader can't get traction.

For example, assume that an existing 1½:1 slope must be cut back 15 feet. The dozer will start at the top of the slope back 15 feet, dozing the excess dirt out each end or over the existing slope.

As the dozer cuts the new bench 15 feet wider, it trims the new slope with the slope bar. This works well for recutting existing slopes because the dozer is the only equipment needed on the slope. The excess dirt that's shoved over the existing slope can be loaded into trucks at the bottom with a loader or hoe.

The operator should run in the lowest gear and make a small cut on each pass. He may leave some boulders sticking out of the slope. Some of these can be left in place and others must be plucked out individually. Usually a 2 foot plus or minus tolerance is permitted on a rock slope, depending on the rock-to-soil ratio and size of rock. After a few passes the grade setter can check the larger boulders sticking out of the slope. He'll tell the operator which boulders are within tolerance and which should be plucked out. After each slope pass, the operator should level and smooth out the bottom grade the dozer is working from. A rough working surface will result in a rough cut slope.

Look at Figure 23-2. This mountain slope was cut with a dozer and sloped with a slope bar. Notice several large rock outcroppings were left protruding from the slope.

Loading a Scraper in Rock

Loading a scraper in a rocky cut is a delicate procedure. Alertness is essential. Be careful not to spin the tires. It can ruin the tires and cost an operator his job. Apply little or no power while being pushed. Let the pushing dozer do the work. This will avoid any tire spin. If you hit a hard spot and the scraper stops or nearly stops, pull the bowl up just a little until the boulder that was causing the trouble has passed. Then let the bowl down slowly until the push tractor has to strain slightly to keep going. You can tell when the push tractor's engine is starting to lug down by watching the tractor's exhaust stack. If you see smoke start to increase, lift the scraper bowl. The push tractor is starting to lug.

In Figure 23-3, notice that the scraper is loaded to overflowing yet no smoke is coming from the dozer's or the scraper's stack. They're not lugging because the operator is taking even cuts.

Mountain slope cut with a dozer and slope bar
Figure 23-2

A dozer that's working efficiently, not lugging
Figure 23-3

In loading rock the scraper operator must raise and lower the scraper bowl constantly. The cutting edge frequently catches on boulders that will not budge. Don't try to cut through them. Let the dozer rip them and then pick them up on the next pass. Occasionally the scraper will load a boulder so large that it won't pass under the bowl when the load is dumped. The bowl can't be raised high enough. In this case dump the boulder on the ground, back up slightly to give the scraper a little turning room, and then turn sharply so the bowl passes beside the boulder instead of over it. Be careful to back up slowly, so you don't hit the transmission housing with the boulder.

Always work slowly and carefully in rock. Don't rush. No one can work fast enough to make up for the loss of a tire, or damaged equipment.

Compacting Fill with Rock

Running a compactor on the fill produced on a rocky job is also difficult. It's the compactor operator's job to see that the fill doesn't become so rough that the scrapers can't move over it. He must either find low areas to push the boulders to, or dig a hole for them so they can be covered quickly. Some can be pushed to the outside edge of the fill until they can be covered. But the operator must be careful that they're not left sticking too far out of the slope. If they slide over the edge of the slope, he may not be able to pull them back into the fill.

Observe the same cautions when dozers are working on the fill area. Keep the fill as smooth as possible. Always keep the boulders pushed to the low spots so they can be covered with the smallest effort.

Ripping Rock with a Dozer

When ripping boulders or rock layers with a dozer, it's usually best to work from north to south or south to north. The cracks in the rocks usually run in that direction. In most cases working north to

One ripper rips lava cap
Figure 23-4

south makes ripping easier. If this isn't the case in your area, study the rock layers to determine the best direction to rip.

The lava cap in Figure 23-4 is being ripped 12 to 18 inches at a time. This rock has no cracks, so it's being ripped, then cross-ripped. Only one ripper is used. After each ripping, the ripped rock is dozed up and loaded into trucks to be used as fill in a low area. Sometimes, scrapers are used if the fill area is near. Then you need a grader on the haul road, blading off rocks that fell from the scrapers. Rocks get stuck in the scraper's apron, wedging it partially open and allowing more rock to fall out as it travels along the haul road.

Any time you're operating equipment in rocky areas, follow these rules: *Think before starting.* Make a smooth area to work from as soon as possible. Doing these two things will increase production. Don't try to overpower the boulders.

Be careful working parallel on a rocky slope with a dozer. If both tracks get on a rock cap at the same time, the dozer may slide. Tracks will slide easily on rock and may cause the dozer to turn over. If this should start to happen, hold the brake and pull the friction on the uphill side. Usually this will swing the rear end downhill and point the front uphill, stopping the slide.

If you find yourself in a precarious position on a steep slope with any piece of equipment, lower the dozer blade, scraper bowl, bucket, or blade to the ground. This may keep the equipment stable until a choker line can be attached to keep it from rolling.

Working a dozer in a rocky area may cause sparks as tracks grind over rocks. If there is dry grass or leaves in the working area, a fire may start. Be prepared in case this happens.

Ripping and Compacting Asphalt Road

Rising construction costs have made it popular to rip and recompact existing asphalt roads to create a stabilized base. This chapter will describe how to do it when you only have a cross-section to guide you — no grades have been set by surveyors. It may seem easy to do a job like this with no grades to follow. It's not. With no grades to work to, the crew must establish its own grading plan. Otherwise they'll waste time moving material from one side of the road to the other. Even worse, the cross sections may not conform to the road specifications.

The main concern on this type of job is the thickness of the existing road surface. If the job specifications don't supply this information, you'll have to dig test holes to determine the exact thickness of the asphalt.

On this type of job, the surveyors will set roadside stakes and indicate centerline distances only. No grades will be set. The job specifications will include a typical road section sheet that shows what work is necessary. The grade setter will set the grades as the work proceeds.

Let's look at the steps you'll follow in a typical road job. Assume the road section requires that you widen the road 2 feet on each side with a 2 percent crown from the centerline to the shoulders. You won't do any work on the existing ditches. What's your first step? Choosing the right equipment for the job.

Selecting the Equipment

The thickness of the asphalt will determine what equipment you'll use. If the asphalt is 2 inches thick or less, a Model 14 grader with a rear ripper rack can handle the ripping. You'll need a compactor that weighs at least 38,000 pounds, with pad type drums. If the asphalt is thicker than 2 inches, the compactor should weigh 60,000 pounds or more. Look back to Figure 7-4 in Chapter 7.

A Model 14 grader can rip asphalt up to 4 inches thick and still get good production. Use a Model 16 grader with a rear ripper rack if the asphalt is 5 or 6 inches thick. Figure 24-1 shows a 16G grader. It has much more weight and power than the 12G. It's ideal for heavy ripping and grading jobs. Notice the ripper bar and shanks mounted on the rear. When ripping asphalt, leave down only every other shank, or even less, depending on the thickness. In many cases, only three rippers will be used.

Ripping and Crushing the Asphalt

After the asphalt has been ripped and rolled three or four times with the compactor, use the grader to scrape it into a windrow. This will make crushing the material much easier. The compactor makes a few passes, then the grader operator rolls the asphalt chunks over into another windrow. The compactor again rolls the windrow. Continue this process until the asphalt chunks are no larger than the required maximum size. Then you can spread and roll the material to a smooth surface or use it as fill for widening shoulders.

A powerful 16G grader
Figure 24-1

When the road can't be closed, rip and break the asphalt on only one lane at a time. This allows traffic to pass during construction. On a two lane road, you'll need a flagman to control traffic, letting cars pass in one direction at a time.

When the ripped area has been shaped, watered, and compacted, direct the traffic over it while you continue the same process on the other lane. Keep an equal length ripped down each lane at the end of the shift. After a day's work, try not to run traffic over road sections that have an edge that doesn't match the existing road beside a ripped lane. This could cause a car to go out of control. If it can't be avoided, delineate the edge well so traffic won't try to run on it.

Grading the Job

After the entire length of the job has been ripped, pulverized, compacted, and the shoulder fills are completed, you're ready to begin grading. The best method for balancing the cut and fill areas on a crowded road is fairly simple. Every 50 or 100 feet, have the grade setter shoot the shoulder elevations from one side to the opposite side. If one shoulder edge is 0.40 higher than the other, adjust the grade accordingly. On the side that's 0.40 higher, he'll place a lath indicating a cut of 0.20. On the side that was 0.40 low, he'll set a lath at the shoulder edge indicating a fill of 0.20.

Set grades on the entire length of the job in this manner, splitting the difference in grade from one side to the other. Be sure to check the width from the centerline set by the surveyors. The grader operator then makes his cut on the high side to bring the excess material across to make the fill on the low side. This will leave the shoulders level at each side of the road.

Checking Centerline Subgrade

Now check the centerline. Shoot from the shoulder grade to the centerline. Compute the percent of slope needed for the crown. Suppose the centerline subgrade is 0.20 too high. Follow these steps to cut it to the finished grade. The grade setter indicates a cut of only 0.10 on the lath at the centerline. The grader makes this 0.10 cut, leaving the excess fill on the shoulder. Adding 0.10 of fill to the shoulders on both sides of the road will level the road just as well as if you had trimmed off 0.20 and hauled it away. Now the centerline will be higher than the shoulder, giving the desired 2 percent slope without using a scraper to balance the grade.

If the centerline is low by 0.20, the grade setter should indicate a 0.10 cut at each shoulder. This would give the centerline the fill needed for a 2 percent slope. This is the fastest and simplest method to use when no grades are supplied. Use the same method if the road has a super elevation (sloping in only one direction, with no crown).

Setting Aggregate Grades

Once subgrade has been trimmed and rolled, aggregate grades can be set. The grade setter must set lath at each shoulder edge, measuring up from the subgrade the height of the aggregate desired. Then he places a horizontal line and an arrow pointing to it (a crows foot), indicating the required fill. Unless the road is very wide, setting centerline laths is a waste of time. Traffic and trucks placing the aggregate would knock down the stakes. But if the road *is* wide enough, stake the centerline also.

Keep the aggregate slightly above grade at the road center and leave both shoulders a little low. Calculate the quantity of rock needed every hundred feet, then dump and spread that amount. Once the aggregate has been placed and rolled, the grade setter should run a row of grade hubs down each edge and centerline. These are used for the fine grading. He must dig down to the subgrade material at each hub and measure up from the subgrade the appropriate aggregate thickness. You'll only have to use this method of controlling the aggregate thickness when there are no grades set by the surveyors.

Placing the Aggregate

It's important to leave the aggregate near the centerline high because car and truck traffic will cause the rock to separate from the fine material. This is usually called *raveling*. Water the aggregate well, cut the excess from the centerline, and spread it to the shoulders which were left low. This will mix the rock and fine material together, leaving a smooth, firm base. At the end of each shift, all the aggregate trimmed and finished should be oiled and sanded so the traffic won't ravel it again. Be sure to put enough sand over the oil so the cars won't fling oil on the fenders. When it's time to pave, sweep off any excess sand and you have a road base ready for paving.

After the road has been paved, shape the aggregate shoulders. Then place the dikes, fog seal the road if required, and do the striping.

Pavement Removal

Removing an old road and replacing it with a new surface on the same alignment presents different problems than a job where you're creating a new road alignment. There are two major differences: First, the first layer of material you'll encounter is asphalt rather than soil. Second, if the road can't be closed, you'll have to handle traffic during the job.

Removing the Asphalt

The most efficient way to remove the asphalt is to use scrapers. If the asphalt is 4 inches thick or less, you can use a scraper with ripper teeth. The scraper should have at least a 20-yard capacity bowl. It can be equipped with a paddle wheel, although many foremen believe that asphalt is hard on a paddle wheel scraper. They'll use only open bowled scrapers, a practice that I feel is overly cautious. A paddle wheel scraper with good teeth will pull up the asphalt and stack it neatly in the bowl. As long as you keep the paddles at a slow speed, this will put little stress on the paddle unit.

Assume that the pavement on a 36-foot wide road has to be removed while the existing traffic is allowed to pass. Follow these steps:

1) Make a cut across the road at each end of the job with an asphalt saw, jackhammer, cutting wheel, or hydrahammer.

2) Have the scrapers start at the right or left of the centerline, depending on which side is to be removed first. If the pavement is 4 inches thick, you may need a push dozer or grader behind the scraper to reduce strain on the transmission and tires, especially if the asphalt is cold. If there are only 2 inches of asphalt, no pushing is needed unless you're using open bowl scrapers.

The operator should set the bowl down so the teeth sit just below the bottom of the asphalt. The remainder of the cutting edge, which is higher than the teeth, rides just at the top of the asphalt. This way the asphalt will load easily. The scraper will cut a strip only as wide as the ripper teeth, approximately 6 feet. It won't   the full 12 feet of the bowl width.

3) The next pass the scrapers make will be at the outside edge of the road. After the outside edge and area near the centerline have been cut, only an area down the center remains. Remove this in the final pass. Be careful not to overfill the scraper bowl. A chunk of asphalt could fall from the top of the bowl and roll into the traffic lane.

4) Now the grader can level and smooth the road base so that traffic can be diverted to that side while the remaining side of asphalt is removed in the same manner.

Don't rip the asphalt before the scrapers remove it because ripping makes the asphalt stack up in front of the bowl. The asphalt will slide ahead of the scraper, making it harder to load. Ripping

A track hoe rips and loads asphalt
Figure 25-1

the asphalt before the scraper starts loading makes it impossible to use a paddle wheel scraper without damaging the paddle unit. You'll have to use an open bowl scraper or loader unless the asphalt is very thin or has been smashed into small chunks.

If possible, put off all pavement removal and cutting until midday. The sun warms the asphalt and makes it easier to cut and load.

When you're working with very thick asphalt or in narrow areas, a hoe may be the best equipment for asphalt removal. In Figure 25-1, a track hoe is pulling up chunks of asphalt 10 inches thick and loading them in trucks to be hauled away.

Loading the Asphalt Chunks

In most cases where the asphalt is thicker than 4 inches, it's advisable to remove it with a loader. You can use either a rubber tire

Loading asphalt with a track loader
Figure 25-2

loader or track loader. And when you're using a loader, it's usually best to rip the asphalt before loading. This will roughly double the rate of production.

A track loader is faster in loose base rock because it has a shorter turning radius and better traction than a rubber tire loader. Good traction is needed to penetrate below the chunks of asphalt when loading. Look at Figure 25-2. The track loader is getting a bucket load of aggregate base and a 2-inch asphalt concrete section with little effort.

When loading trucks with a loader, there are three rules to speed production. The first is to load from a level area. The second is to approach the truck squarely when dumping. The third is to try to park the truck bed 90 degrees from your loading position. The operator must stay far enough from the truck so that the load will be dumped in the center of the bed. See Figure 25-3. On a small

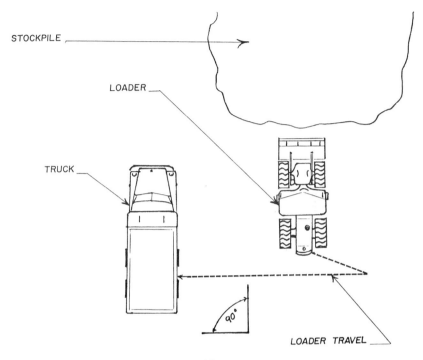

STOCKPILE

LOADER

TRUCK

90°

LOADER TRAVEL

PROPER LOADING ANGLE 90°

Proper loading angle is 90 degrees
Figure 25-3

job where the cost of moving in a ripper cat would be too high, the loader can break up the asphalt as it loads.

Schedule Enough Trucks
Drive and time the route the trucks use from the load area to the dump area before the loading starts. Determine the time required for a truck to be loaded, travel to the dump site, and return. Once you know the travel time, order out enough trucks so the loader isn't kept waiting.

Loading with a Backhoe

When loading chunks of asphalt into a dump truck with a backhoe, make sure the asphalt chunk is clamped well between the bucket and hoe arm. If it's not clamped well, it can slip out and fall when the hoe arm is raised to dump into the truck bed. As the asphalt chunk slips free, it will usually fall toward the hoe. On a rubber tire hoe, there's a good chance it will land on the outrigger arm, breaking off hydraulic connections.

If the hoe has outriggers, the operator, besides clamping the asphalt chunk well, should raise the hoe tractor as high as possible. This is done by extending the outriggers down as far as possible, then lowering the front bucket. This raises the front of the tractor until it's level with the back of the tractor. When a small hoe is set up as high as possible, it reduces the steep angle the hoe arm must raise to dump into the truck bed. This should help keep the asphalt chunks from slipping out of the bucket.

Asphaltic Concrete Paving

This chapter will introduce you to the various methods and equipment used in placing and finishing asphaltic concrete. Self-propelled paving machines are the primary paving tool today. They're used on nearly all larger jobs where a specification quality surface is required.

Self-propelled Paving Machines

There are two ways for self-propelled paving machines to receive the asphalt. First, end dump trucks can dump the mix into the paving machine hopper. Second, a pickup machine can paddle up asphalt which was dumped from bottom dump trucks. The pickup machine works like a self-loading scraper. It picks up the asphalt from the windrow that was dumped on the ground. The paddles carry it to the top of the machine and flip it into the hopper at the front of the paving machine.

Pickup machine loads A.C. into hopper
Figure 26-1

Figure 26-1 shows a paver moving along at an uninterrupted pace as a pickup machine paddles the windrow of asphaltic concrete into the hopper. At the extreme left of the picture, a truck has just pulled ahead of the paver and the dump man is hand dumping another windrow of A.C.

Track and Rubber Tire Pavers

There are two styles of paving machines, track pavers and rubber tire pavers. The track paver is excellent for paving on aggregate base, especially if the aggregate is a little loose. The track paver has much better traction than a rubber tire paver, which will spin the tires occasionally when the base is loose, and may actually get stuck in the aggregate. If the aggregate is firm and if the rubber tire paver doesn't get overloaded, it will do a fine job. The rubber tire paver is excellent for overlays where it can run on a hard surface.

This hopper receives the asphalt
Figure 26-2

The Hopper and Conveyor Belts

From the hopper, the mix is carried to the back of the paver by the two conveyor belts. Figure 26-2 shows the hopper at the front of the paver. The two conveyors deliver asphalt to the back of the paver and dump it on the ground ahead of the screed. Notice the left conveyor gate has been lowered to restrict the amount of asphalt on that side. This works well on a leveling pass when one side is taking less asphalt. Both conveyors can still run automatically without the operator manually controlling one side. You can see a rubber belt that has been added across the front of the hopper so when the truck dumps, the asphalt won't spill on the ground. The two rollers at the bottom of this picture are placed against the truck tires as the paver pushes the truck ahead.

Paver leaves a smooth, partly-compacted mat
Figure 26-3

The Screed Augers

Two screed augers spread the asphalt concrete the entire length of the screed. The screed strikes off the asphalt concrete at the set depth. The asphalt concrete passes under the screed and comes out of the back of the screed smooth and partially compacted. Look at Figure 26-3. The paver is moving along leaving a smooth, partly-compacted mat behind. The paver has a screed extension on each side and the hydraulic screed (shotgun) is extended 18 inches on the left side. A double drum vibratory breakdown roller is working closely with the paver as the tandem finish roller makes a pass over mats previously laid.

Two sensors monitor the amount of asphalt that's dumped to the augers from the conveyor. These sensors work independently of each other. If they detect that the augers are becoming overloaded with asphalt, they immediately shut the conveyor off

Hydraulic screed extension with control sensor
Figure 26-4

until the augers can distribute the accumulated material. The automatic sensors can also be turned off so the mix supplied to the augers can be controlled manually. The operator then controls the conveyor delivery rate. Once the augers have distributed the asphalt the entire length of the screed, the screed both levels the asphalt at the depth set by the screed man and compacts the mix with a vibrating, tamping motion.

The Hydraulic Screed Extension

Most pavers are equipped with a hydraulic screed extension which can be extended out 4 feet when needed. In Figure 26-4, you can see the extension extended 2 feet on the right side of the paver. The arm on the outside of the hydraulic screed extension has the automatic screed control sensor on the end. Notice the cord running to it.

Expandable paver paves a 32-foot pass
Figure 26-5

The extension is used for a road area that's wider than the standard width for a short distance. The hydraulic screed extension can be extended to catch the extra width without interrupting the operation. Then the area can be tapered back to the standard pass width by retracting the extension. The extension screed leaves the mat higher than the rest of the screed because it doesn't have the vibrating compacting action. The mix is left higher to allow for the extra compression when this area is rolled.

Expandable Pavers

There are many models and sizes of paving machines. Some models will lay pavement more than 20 feet wide. They come with extending tubes and have vibrating screeds the width of the extension. You can recognize these pavers by the chrome tubes that run across the back of the paver. They slide out as the screed is extended. Figure 26-5 shows an expandable paver that is paving a 32-foot pass crowned at the centerline. Both the bolt-on extension and the slide-out extension, if extended several feet, must have auger extensions bolted on to get the asphalt out to the extensions.

When the paver is extended past 18 feet, the asphalt must be augered further to reach the extensions at the outer edge. The operator may need to slow his travel speed by dropping to a lower gear if a deep section is being paved. That way the augers can catch up with the material requirement of the screed. If a 10- or 12-foot pass is being paved, the augers can usually supply enough material to the outer edges of the screed to permit a higher travel speed.

Using a Pickup Machine

Using a pickup machine to move hot mix to the hopper will increase the hourly tonnage handled substantially. The big advantage in using a pickup machine is that the paver almost never stops. When using end dump trucks, the paver must stop every time the truck is empty so that a full truck can back up to the paver and raise the bed to dump.

The paving machine supplies the power that pushes the truck ahead when end dump trucks are used. The paver also propels the pickup machine which is attached. The pickup machine is only effective on long passes when the paving machine is not required to start a new pass frequently. When a pickup machine is used, the asphalt is dumped on the ground using bottom dump trucks rather than lift bed end dump trucks.

If you're using a pickup machine, adjust it periodically so that it scrapes up as much asphalt concrete as possible without cutting into the subgrade.

Setting the String Lines

A string line guide must be set on the ground before paving begins. The string is set 6 inches off of the centerline so it doesn't get paved over. This way, the paving machine operator can keep the string line in sight at all times. On a wide road, the paving foreman may elect to set a string line at the shoulder edge of the pavement, again using a 6-inch offset.

Usually no string line is needed when paving a parking lot because curbs serve as a reference line. If extruded curb is to be placed on the pavement after the parking lot is paved, then the edge of pavement line around planters and the perimeter should be marked with paint by the grading crew. It's a time-consuming job, but they can do it quicker because they're more familiar with the plans, having just graded the aggregate base.

Planning the Passes

Let's assume the roadway being paved is 64 feet wide, 32 feet each side of the crown down the centerline. Included in the 64-foot width is an 8-foot shoulder on each side with a different percent of slope than the roadway. Here's the best procedure to pave a road width of this size. Set a string line at the centerline with a 6-inch offset away from the area being paved. Once the string is set, pave two 12-foot strips starting from the centerline. The two 12-foot passes total 24 feet, leaving only an 8-foot section for the shoulder on the first half of the road. But in this case, since the shoulder slope is different, the foreman should pave the 8-foot shoulder before the second 12-foot pass. If the shoulder isn't paved first, the paver would have to run on the last roadway paving pass to pave the shoulder.

It's a good practice to run another string line at the shoulder edge of pavement before paving the 8-foot shoulder pass. The same procedure will be used on the other half of the road with one exception. No string will be needed at the centerline because the pavement there is placed flush against the first pass.

If the shoulder grade is the same thickness and percentage of slope as the road section, most agencies will allow you to make two 16-foot passes for the 32 feet. Most agencies will allow you to place a crown in the screed and pave a 16-foot pass down the centerline, leaving 8 feet in each lane. Then, with the crown taken out of the screed, two 12-foot passes paved on each side of the centerline strip would yield the 64-foot total required. The width you choose to use depends on the width the paver can pave most efficiently, and, if you're working on an existing road, the traffic conditions.

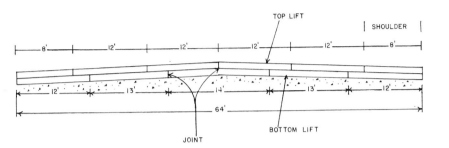

Stagger the joints on two lifts of asphalt
Figure 26-6

Remember, if the shoulders have a greater slope than the road section, they must be paved separately, and before the last road pass along the shoulder. Some agencies insist that pavement joints between passes be at the edge of the lanes where the traffic stripe will be and not in the traveled lane. Read the specifications so you know the requirements that must be met.

Some pavers are designed to pave a shoulder slope simultaneously with the travel lane at a different slope percentage. If you're using one of these machines, you may be able to pave the road and shoulder in two 16-foot passes.

If two lifts of asphalt are required, the joints of the two lifts should be staggered. If the top joint should crack with age, the bottom lift still remains to hold the surface intact. See Figure 26-6.

Preparations Before Paving

Some preparation must be done before a paving machine is ready for paving. Allow time for this. Under average weather conditions the paving screed must be warmed up for 20 minutes. Paving machines have a pump which blows diesel oil spray into the screed tunnel, where electric fire starters ignite the diesel fuel. While the fire is burning in the screed tunnel and the engine is running, the operator should check the condition of the machine. Any area on the paving machine that comes in contact with asphalt must be sprayed with a light coat of diesel. This will prevent the asphalt from sticking to cold parts.

If a pickup machine is used, it should be started, checked for worn or loose parts, and sprayed with diesel fuel. The paving machine is equipped with a pump hose and spray nozzle for spraying diesel on any spot where needed. All of the bearings on the pickup machine and the paving machine must be greased and checked for wear. The flight chains that motivate the conveyor on the paving machine must be checked to be sure they haven't loosened.

Planning the Dump

If the asphalt delivery trucks are scheduled correctly, the paver can continue on without stopping. When a pickup machine is being used, the paver operator should try to adjust the speed of the paver in accord with the number of truck loads of mix available ahead.

The dump man must be careful to see that the right amount of mix is dumped ahead of the paving machine. He must know the distance each truck load of asphalt is to be spread. Assume, for example, that 25-ton bottom dump trucks are used. Each trailer will carry 12½ tons. And assume that we have computed the width and thickness of the asphalt mat being paved. We find that it will take the paving machine 180 feet to use the 25 tons dumped.

Checking the Truck's Spreading Characteristics
If you are not familiar with the bottom dump truck spreading characteristics, set the chains on each trailer so the bottom gates will open only 12 inches. Dump the back trailer. Then measure the distance it took to unload. If the distance it took to spread one trailer was more than 90 feet, the mix was dumped too light. If the distance was shorter than 90 feet, it was dumped too heavy.

If it was dumped too light, have the truck pull out and come around again. The truck should straddle the windrow just dumped. Reset the chains on both trailers to dump a little heavier.

A good job by the dump man
Figure 26-7

Hand operate the trailer gates to finish dumping the remainder of the mix in the first 90 feet. Once the 90 foot mark has been reached, start dumping the front trailer. Always remember that the complete load must be dumped over no more than 180 feet.

If the first trailer dumped too heavy (say it spread only 70 feet), adjust the flow control chains on both trailers a little shorter. Move the truck ahead 20 feet so the second trailer will start dumping at the 90 foot mark. Never dump more than one load in 180 feet. It's much easier to dump a little more asphalt later than to get rid of an excess. A dump man with a few years of experience will usually hand dump each trailer. He should be able to judge exactly how much is needed without measuring the distance dumped. The dump man controlling the dump in Figure 26-7 has done an excellent job. The even, continuous windrow means a good production rate.

If the bottom dump trucks are going to be hand dumped, which means the dump man will control the gates with a lever located on the left side towards the rear of each trailer, the front trailer should be dumped first. Once the front trailer is empty, the dump man will stand at the spot where dumping was finished until the back trailer gets to him. Then he will continue dumping the second trailer. Dumping the front trailer first is possible only on level ground with a good firm base or when dumping downhill. If the grade is loose or if the truck is dumping on an incline, the truck will lose traction when the front trailer is empty.

If the back trailer is dumped first, the truck has to stop and back up until the front trailer reaches the spot where the back trailer ran out of mix. You always want to dump a continuous windrow.

Operating the Paver

The number of men required with the paver will vary from four to seven, including the operator. More men are needed for raking and shoveling if there are cul-de-sacs or tight curves in parking lots to pave. If a pickup machine is used, another man is needed to handle the dumping. If you're paving where traffic is a problem, two or more workers may be needed for flagging traffic.

Watching the Mix

As paving progresses, the foreman should watch the temperature, texture and oil content of the asphalt closely. If the mix doesn't have enough fine material, it will look rocky and coarse as it leaves the screed. When the mix is too hot, it will smoke more than usual and may even look brownish. The mix won't roll well under the first roller and will tend to leave a pebbled surface. When the mix is too cool, you can tell as it's dumped from the truck. It will be slow sliding into the hopper and may even be a little lumpy.

Judge the amount of oil by the shine of the mix. A dull surface means it needs more oil and a very shiny look means it has too much oil. Call the asphalt plant if any of these problems occur, so the problem will be corrected. It's good practice to have an asphalt thermometer on hand to check the asphalt temperature regularly while paving.

Open graded asphalt is usually spread at a temperature of 200 to 250 degrees. Regular asphalt is usually spread at 250 to 325 degrees. Most job specifications will not allow asphalt to be placed when the air temperature is 40 degrees or less. Don't place open graded asphalt when air temperature drops below 60 degrees. Since open graded asphalt has no fines, it doesn't retain heat as well.

In cold weather or on long hauls, make sure the asphalt is covered with tarps during the trip from the plant to the job. In very hot weather, and when traffic must be diverted over the mat just paved so the remainder of the street can be paved, the inspector may require that a water truck spray the asphalt mat to cool it. Traffic can damage a hot mat of asphalt concrete on a turn or in an area where the traffic must stop and start quickly.

Controlling the Percentage of Slope

Put down the first strip or mat of asphaltic concrete at the centerline. Lay it with the automatic screed height control if the paver is equipped with one. Set the screed control sensor to drag on the subgrade surface at the centerline. Then dial the percentage of slope of the road into the automatic slope control. On the second and any following passes, the screed control sensor drags on the top of the asphalt of the first pass. Figure 26-8 shows the screed control sensor on the asphalt.

Check the slope percentage with the slope level indicator at the back of the paver just above the screed. You can see the level near the top of the photo in Figure 26-9. At the bottom center, you can also see a nut next to the chain. Turn the nut to roll the chain to create a crown in the screed.

Check the indicator closely to be sure it shows the same percentage of slope as is set on the automatic slope control. The slope indicator is very important when the percentage of slope in the road is changing. For example, assume the percentage was changing from a minus 3% to a plus 3%. As the change is dialed into the automatic slope control, the screed man should watch the slope

Automatic screed control sensor
Figure 26-8

Slope indicator level
Figure 26-9

percentage indicator to make sure the screed is reacting properly. If the paver has grade control skids dragging on each side, they will control the paver to exactly follow the subgrade or existing pavement.

Adjusting the screed— When starting a pass with the paving machine, place blocks under the screed on each side so that it's off the ground approximately the thickness of the asphalt desired. Screw the screed elevation adjustment up or down until you feel a slight ease in turning the screed handle on each side. Then turn the screed adjustment so that most of the weight is taken off the blocks. This should set the screed so that it won't drop as the paver pulls ahead and the screed leaves the blocks.

The first few feet traveled on a new pass may require some quick adjustments, even when using screed control sensors. It's good practice to use two men on the screed for the first few feet. After about 50 feet, you usually only need one man. When you're tying into an existing pavement, set a lath under each screed edge to hold the new pavement level up enough to allow for shrinkage that will occur when it's rolled.

The screed man should measure the asphalt thickness several times in the first several feet of each new pass and thereafter as he feels it's needed. The stab rod used to check the depth or thickness of the asphalt can be made of a washer welded to a sleeve as shown in Figure 26-10.

Correcting the screed to allow for compaction— The screed does some compacting of the asphalt. The asphalt mat will be compressed further when the rollers have finished smoothing the surface. This shrinkage must be figured into the depth of the asphalt mat as it passes under the screed. You don't want the finished asphalt mat to be thinner than called for in the specifications. Figure 26-11 shows the depths needed at the screed to achieve the finished depth required after rolling.

Adjusting the screed for crowns and swales— There may be times when the paver must straddle a crown in the road or may be re-

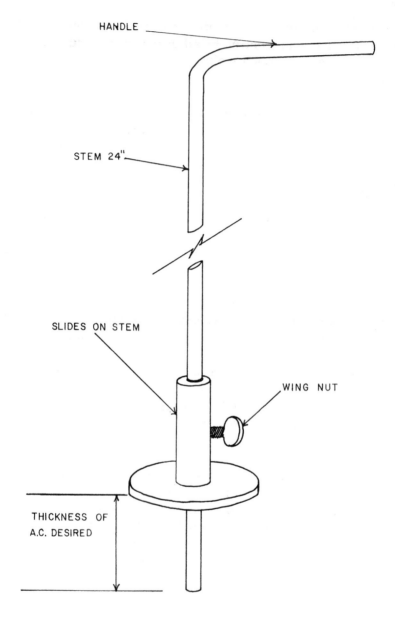

HANDLE

STEM 24"

SLIDES ON STEM

WING NUT

THICKNESS OF
A.C. DESIRED

Stab rod for checking depth of asphalt
Figure 26-10

ASPHALT CONCRETE SCREED SETTING (DEPTH)	ASPHALT CONCRETE AFTER ROLLING (DEPTH)
1 3/16"	1"
1 3/4"	1 1/2"
2 3/8"	2"
3"	2 1/2"
3 5/8"	3"
4 1/4"	3 1/2"
4 7/8"	4"
5 1/2"	4 1/2"
6 1/8"	5"

DEPTHS MAY VARY SLIGHTLY WITH VARIOUS AGGREGATES

Screed settings adjusted for compaction
Figure 26-11

quired to pave a swale. The screed on the paving machine can be adjusted for a crown or swale pass. All paving machines aren't the same but they all work on the same principle. Above the screed there will be a nut at the center of the paver. Refer again to Figure 26-9. The screed is hinged at the center of the paver and will pivot on its hinges as the nut is turned, one way for a crowned mat and the other way for a swaled mat. Use a string line stretched across the screed bottom to check the amount of crown or swale that has been set. When using the automatic screed control, be sure that the manual screed adjustment wheels are screwed to the center before the automatic screed takes over.

Stopping the Conveyors

The paving machine operator must stop the conveyors each time he comes to the end of a paving pass. He must judge the material and distance so the screed will be empty when the end is reached. It's best to stop the conveyors a little early. More mix can be conveyed to the screed if needed.

Any excess material ahead of the screed at the end of the pass will be left on the ground when the screed is lifted, It must be shoveled back into the machine or spread on the grade where the next pass will cover it. If a pickup machine is being used, however, the paver can be turned around and the excess asphalt can be paddled up with the pickup machine.

Paving Grade Changes

Paving on and off ramps on highways is usually difficult because the grade changes rapidly from one slope to another. There will usually be a long tapered area to pave and perhaps a narrow shoulder at a different slope than the traveled lane.

With automatic slope control— If you're paving a slope that changes from a plus to a minus grade, and if the paving machine has an automatic sensor which senses the grade from a string line or wire, the operator can change the automatic slope control smoothly at the station where the grade changes before he reaches that station. The grade setter should write the new slope on a 3 x 5-inch card and tack the card on a lath placed before the change station.

Without automatic slope control— Grade changes are more difficult with paving machines not equipped with automatic slope control. It takes two men to operate the screed. Each man must watch both the thickness of the asphalt being spread on his side and the slope level. The operator should run the paving machine slow enough so that the men on the screed can react to the grade changes. Again, use 3 x 5-inch cards to warn of grade changes so the paving machine operator can let the screed men know in advance of grade changes. If there are shoulders to be paved, they should be paved before the traffic lane has been paved, if possible.

Paving tapered and curved areas— When paving tapered areas, it's usually best to stop the full pass just short of where the taper begins and then pave the taper first. Once the taper is paved, continue the full pass over part of the taper. This will make a smooth transition and both areas can be rolled together while still hot.

If a pickup machine is used on short radius turns, you can usually paddle enough asphalt from the traffic lane windrow to complete the curve. If more mix is needed to finish the radius, paddle more asphalt from the traffic lane windrow. The same procedure should be used on short tapers. The dump man must be careful with his dumping when a taper or radius is involved. The normal dump pattern might leave material in the way for pulling a taper or radius. You might need to hold the trucks up until the taper or radius has been paved before you finish the traffic lane windrow.

If there are smaller areas where the paving machine must pick up and reset frequently, disconnect the pickup machine. Then the paver can't use asphalt dumped on the ground in a windrow. You need end dump trucks to dump mix into the hopper of the paving machine. When the truck has backed up to the paver and lifted the loaded bed, the truck is in neutral and the paving machine pushes the truck ahead as it paves.

Paving cul-de-sacs and parking lots— When paving subdivision streets, pave cul-de-sacs with their short radius curves first before making the main pass. When paving a cul-de-sac or other difficult areas, you'll probably need two men on the screed. Make the first pass at the outside edge, moving in toward the center with each additional pass. Then use the same procedure on the opposite side. Figure 26-12 shows the steps to use for paving cul-de-sacs. Start with step 1 and proceed through step 10. Pave intersection curves before the traffic lane is paved. You can eliminate steps 2 and 6 when using a paver with the tube-type expandable screed.

A good deal of raking and shoveling is required behind the paving machine in areas such as cul-de-sacs. You can pave cul-de-sacs

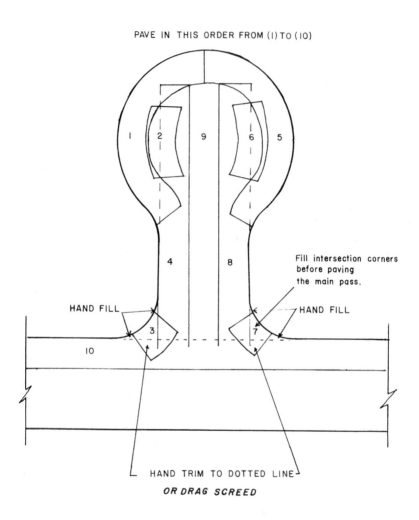

Paving cul-de-sacs
Figure 26-12

with a pickup machine on the paver. To do this, build a ramp of dirt over the back of the curb and sidewalk so the trucks can drive over them after they dump. If a deep section of asphalt is being paved, a small asphalt ramp might be needed on the street side of the curb.

You can pave around islands in parking lots with a paving machine, although some hand work is involved. When there are several islands in a small area, have a small tractor with a bottom grading blade and back drag fill all the corners just ahead of the paving machine. This will free the paving machine for longer passes and speed up the operation.

There are times when the paver must pave areas so short or curved that a truck can't maneuver with the paver. The paver must get a full hopper of asphalt concrete, leave the truck, and lay down the short pass on the curved area or in a corner. If more mix is needed, the paver must return to the truck for more mix. If there are many of these areas, it may be faster to dump the asphalt concrete on the ground and have it carried to the hopper with a loader. This will free the truck for another load, so fewer trucks will be needed.

Controlling the Screed

The screed man should develop a feel for the particular paving machine he's working on. Each screed has its own characteristics. A screed man should be used on the same paving machine if possible. When paving a second pass parallel to the first pass, be sure that the screed butts against the first pass. A small gap between the two passes can be filled by raking material into the gap. However, this fill will sink a little after the pavement has been traveled on for a few months. Butting passes up tight makes a better joint and saves some hand raking.

Correcting an Overloaded Paver

If you're using a pickup machine and the paving machine becomes overloaded with asphalt because the dump man dumped too heavy, there are two things you can do. If the paving machine operator recognizes the problem soon enough, he can have the screed man raise the screed slightly until the danger of being overloaded has passed. If the grade tolerance is such that the screed can't be raised, the paving machine must be stopped and the extra asphalt removed to some convenient location.

The inspector may allow the paving machine to move ahead and spread a 1/2-inch thick layer of asphalt well ahead of the paving operation until the excess has been used. When the dump man gets to the area where the 1/2 inch of mix was spread, he must compensate by dumping less. On some jobs there will be an area available where the excess asphalt concrete can be piled before being hauled off after the job is complete. But this is expensive. Avoid it if possible.

Dealing with Paver Breakdowns

Occasionally, the paving machine will have a major breakdown that can't be repaired before the windrow of asphalt concrete has cooled. If no other paving machine is available to take over, use a grader to spread the mix very thin so it can be overlaid when the paving machine has been repaired.

If the paving machine should break down for two or three hours, the mix in the paver will be cold enough to break a flight chain or cause other damage to the paver. Once the paver is repaired, raise the screed and pull away from the mat of pavement. If a small amount of mix is in the hopper, shovel it out. Then spread out the pile of mix left on the ground after raising the screed, either with a shovel or with the screed. This may take more than one attempt. If it has turned to chunks, throw them to the side of the road. When a large amount of mix was in the hopper, it may have stayed warm enough to run through the machine. Try to run the conveyor slowly. If the engine lugs, stop the conveyor. Trying to force mix through the conveyor will usually break a flight chain.

Coordinating with Asphalt Trucks

It's important to keep the asphalt truck drivers informed of the paving pattern, especially when several streets are being paved. If there are 20 trucks hauling to the paver, it doesn't take many confused truck drivers to obstruct the progress of the paving crew. This is especially true of bottom dump trucks because they can't back up very far without jackknifing. Keep a smooth flow of trucks in and out of the project by making up a map of the paving pattern. Each driver should receive a copy at the asphalt plant. This will eliminate most of the truck control problems and increase

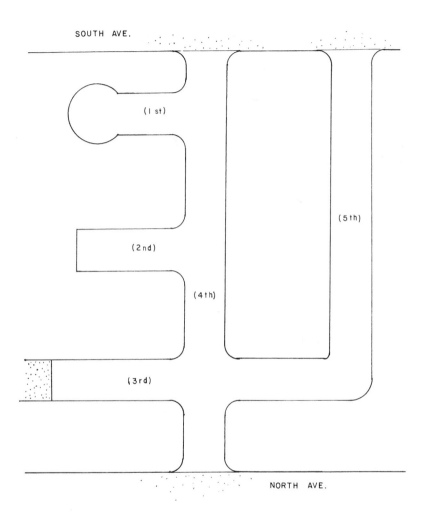

SOUTH AVE.

(1 st)

(2 nd)

(3 rd)

(4 th)

(5 th)

NORTH AVE.

Map for asphalt truck drivers
Figure 26-13

production. Figure 26-13 is an example of such a map. The areas to be paved are marked in order, from the first to the fifth. If the foreman wants to deviate from the map paving plan, most truckers have CB radios and can pass the change to the other truckers over the CB.

Hydraulic arm holds spreader box to truck
Figure 26-14

Paving with a Spreader Box

Paving with a spreader box is essentially the same as using a paving
machine except for these three differences: The spreader box isn't
self-propelled, and it usually doesn't have a vibrating screed, or a
conveyor. The screed adjustments work essentially the same way
but the spreader box has no automatic screed control system.
There's usually no screed tunnel to be warmed up in a spreader
box. As soon as asphalt concrete slides under the screed on the
first pass, stop the spreader box for five minutes and let the asphalt
concrete heat the screed.

The spreader box is pulled by the truck that's dumping the mix
in it. It's attached to the truck by arms that extend into the wheel
wells of the truck's back wheels. See Figure 26-14. These arms
have rollers and are pulled tight against the wheel wells with a
hydraulic jack-type pump. The spreader box rolls on wheels or

tracks. In Figure 26-14 you can see the arm with the roller holding the box to the truck. Notice the small tires of the spreader box at the back of the truck tire.

When each pass has been finished, the box must be lifted by the truck and moved to the next area for the second pass. This is done by hooking a chain from the back of the spreader box to the top of the tailgate of the truck when the truck bed is raised. When the bed is lowered, the spreader box is lifted off the ground. The hydraulic arms hooked to the truck's wheels stay hooked when lifting and moving the box for the second pass.

Most spreader boxes have a hydraulic operated gate that can be shut to hold the mix left in the box. This keeps the mix in the box when it's lifted. Once the bed of the truck has been lowered and the box lifted, the truck is free to move to the point where the next pass begins.

To start the next pass, the truck bed is raised until the spreader box is on the ground and there is slack in the chain. The box can then be unhooked and the gate holding the asphalt opened. The spreader box is again ready to go. Mix is dumped into the spreader box and paving is resumed. Chains on the tailgate of the truck must be adjusted so the tailgate opens only about 15 inches. This will keep the asphalt from dumping too fast and overfilling the spreader box.

Now look at Figure 26-15. There's a screed man with his hands on the screed control handles on each side of the spreader box. The white tank contains propane fuel for the two small engines that run the hydraulics. This box also has a hydraulic screed extension on each side. Notice the rough texture of the asphalt on the far side. This is the asphalt laid by the 1-foot extension being used on that side.

Controlling the Mix and Speed

Use hand signals to the truck driver to control the amount of mix dumped into the box and the speed the box is being pulled. Most truck drivers have a tendency to pull the spreader box too fast. The speed must be kept slow and even. The screed man needs time to react to the grade changes and check the thickness of the mat. A

Spreader box with two screed men
Figure 26-15

steady speed is required to help control mat thickness. If the screed
man makes no screed adjustments when the truck speeds up, the
mat will get thinner. If the truck slows down, the mat will get
thicker.

The raker must keep up with the spreader box. If he can't keep
up, the box must be slowed down or a second raker must be used.

When a spreader box is used, leave the asphalt mat approximately 3/16 to 1/4 inch thicker than when using a paving
machine. The extra thickness is required when the spreader box
doesn't have a vibrating screed.

At the end of each shift, all asphalt must be removed from the
box or paving machine. Every surface the asphalt concrete has
come in contact with must be sprayed with diesel oil. Give special
attention to moving parts.

Coordinating the Trucks

The number of trucks needed for a paving spread depends on the distance they must travel from the plant to the project. The size of the area being paved and the number of cul-de-sacs, short stub streets, and islands will determine the tonnage that a paving crew is able to put down in a day. A parking lot job with several islands to be paved around where end dump trucks are used might require 120 to 150 tons per hour. If a long section of road is being paved, a pickup machine is used, and bottom dump trucks are hauling the asphalt, the tonnage could range from 250 to 450 tons per hour. Experienced superintendents will usually be able to look at a job and determine the daily tonnage. However, it's not uncommon to have to add more trucks or reduce the delivery rate once the paving has begun.

If you're in doubt as to how many tons are needed for a shift, seek advice from the estimator or the paving machine operator. Experienced operators usually know how much area they can pave in a day. And you must know the driving time between the plant and the job site before you can schedule any trucks. After estimating the tonnage needed and the travel time from the plant to the job and back again, you should be able to determine the number of trucks that must be used.

Always use your high estimate when scheduling trucks. Once the paving begins, it's easier to lay off one truck than to add one truck. It's cheaper to have a truck stand by until an adjustment can be made than to have the paving machine and crew waiting for trucks.

When the plant supplying the asphalt is several miles from the job, there may be many trucks on the road at all times. Keep this in mind when coming to the end of the job or at the end of the day. If there are 15 trucks hauling asphalt, and only 15 more loads are needed, the truck drivers must be told not to return after they dump. If load numbers are marked on each scale ticket, the plant can be called and told to stop loading at a given load number.

It's easiest to have the plant control the load cut-off point. But to do this, someone must keep ahead of the situation and notify the plant before the last load required leaves the plant. Some asphalt plants hold prepared mix in a hopper. If this is the case, plant personnel must know in advance of the shut-off time, tonnage, or final load number so the hopper can be emptied before the shift is over. At the start of the day, check with the plant foreman and ask how much advance notice is needed to clear the hopper.

Rolling the Spread

Once the asphalt is spread, the weight of the roller and number of passes will determine the quality of the surface. The number of rollers, the order in which they roll, and the number of passes, are usually spelled out in the job specifications.

When rolling asphalt, always make the first pass on the low side of the asphalt mat. Always roll with the bull wheel first and the tiller wheel trailing. (The bull wheel propels the roller and the tiller wheel steers it.) If the area being rolled is level, start rolling on the edge away from the side of the next pass. If the paver will soon make the second pass, don't roll the last foot of the second pass side. It's much easier to pave up to an edge that has not been rolled. If the weather is warm, as much as one-half hour can elapse before the edge must be tied into and rolled.

Some job specifications for asphalt require that three rollers be used: a double-drum vibratory for breakdown, a pneumatic-tired roller, and a 8-ton tandem. The number of times the mat must be rolled will also be specified.

The initial rolling is done with a double-drum vibrating tandem roller. This is called the breakdown rolling. The vibrating tandem must keep up with the paving machine, never falling far behind. The second rolling is done with the rubber-tired roller and should start when the asphalt mat has cooled slightly, but not cooler than 180 degrees. The third roller is an 8-ton tandem. This is the finish roller.

The number of rollers required depends on local requirements and the size of the job. Some specifications require only a double-drum vibratory breakdown roller and an 8-ton tandem roller to finish, eliminating the need for the rubber-tired roller.

Many small jobs require only an 8-ton tandem for breakdown and finishing. In a parking lot with many islands, use a 2-ton tandem to roll out the creases left by the 8-ton tandem. If open graded asphalt is being paved, a tandem roller weighing 10 tons or less is usually required.

All asphalt rollers must be equipped with fiber mats and water to keep the drum or tires wet. Moist surfaces don't collect asphalt as easily as dry surfaces. Once the tires on the rubber-tired roller get hot, the asphalt won't stick to them so water usually isn't needed.

Applying the Tack Coat

An emulsified asphalt (tack coat) must be sprayed on all edges and surfaces where asphalt concrete joins another surface. The only time this tack coat is not needed is when two hot asphalt joints are being joined together. On an overlay job where a new surface is being put on an existing road, an asphalt tank truck, commonly known as a boot truck, can be used to apply the tack coat. If joints or curbs must be sprayed, an oil pot with a hand sprayer can apply the asphalt emulsions. An oil pot can also be used to spray the tack coat on most small overlays where 200 gallons or less are needed.

The amount of tack coat applied is usually determined by the inspector. It may vary from 0.03 to 0.15 gallon per square yard of surface. Generally 0.05 gallon is applied. The asphalt emulsions will be brownish when sprayed on the base or curb. No asphalt should be placed over or against it until it has had time to turn black. If the weather is hot it will turn black quickly. In cool weather it may take 15 minutes or more.

When spraying a tack coat, be careful not to spray it on anything that won't be paved. Use a piece of plywood to collect overspray when spraying against a wall. Sandblasting may be required to remove tack coat that's accidentally sprayed on a rough surface. A rag and diesel fuel will remove asphalt emulsion from a smooth surface.

Two rakers feather a pass
Figure 26-16

Paving Small Areas, Patches and Trenches

For paving small areas, patches, and trenches, an oil pot and small roller are essential. Patching small areas and trench paving takes an experienced crew to keep the patch from becoming rough and uneven. An experienced raker is needed, especially when tapering the edge over existing pavement. This is called *feathering*. Figure 26-16 shows two rakers feathering a pass just finished into the previous pass on a radius just paved.

There are three steps to follow in feathering:

1) The existing asphalt must be well primed with an asphalt emulsion (tack coat).

2) When the asphalt is applied, all larger rocks must be raked out, leaving mainly fine material.

The roller will roll the feathered edge immediately
Figure 26-17

3) The feathered edge must be rolled quickly before it has time to cool. This will leave a smooth edge. In Figure 26-17, a raker feathers the new street while a roller stands by to roll it as soon as the raker finishes.

The raker must be fast enough to keep the edge from cooling before it can be rolled. A good quality asphalt rake must be used. Otherwise you'll never get a good job. This is about the only time the low side of the mat would not be rolled first. If the cold joint being tied into is on the high side of the mat, run a tandem roller on the cold mat, rolling the cold mat plus approximately two feet of the hot mat. Roll as close behind the raker as possible.

Once this is done, the low side pass can be made, working progressively up to the cold edge. As soon as the rakers have more edge raked, leave the regular rolling pattern and roll the edge again. Hot mix feathered over cold mix must be rolled quickly. Otherwise it will cool and can never be rolled smooth unless more hot mix is placed over it. When making a tie-in or patching in cold or damp weather, a propane burner is useful to warm the existing pavement edge before the oil and asphalt are placed.

The asphalt raker should carry a putty knife with him so he can scrape off asphalt concrete that sticks to the rake. It's important to keep the rake clean, especially when feathering an edge. Diesel oil should be sprayed on the rake regularly.

Placing Asphalt by Hand

When the asphalt concrete is being placed by hand and a smooth surface free of indentations is needed, pave the area in two lifts when possible. The size of the roller depends on the size of the job. To get a final surface when 3 or 4 inches is needed, place the first lift 2 or 3 inches deep and roll it well. Then tack the edges with asphalt emulsion and pave the last 1 or 2 inches. You'll leave a much smoother job if you put down a 1/2 or 1-inch finish course.

Never use asphalt with aggregate larger than 1/2 inch when patching, paving small areas by hand, or when a smooth feathered edge is needed. For skin patches that must be feathered to match existing areas, use 3/8-inch aggregate mix. Always use 1/2-inch asphalt when using a spreader box. In many cases 3/4-inch asphalt concrete will be specified for a top lift put down by a paving machine on road jobs.

A crew that's doing patch work or small paving jobs that involve a lot of hand spreading and tamping of corners should have the following tools: a gas operated plate tamper, hand tamper, spray can, square nose shovels, asphalt rakes, 5-gallon buckets, push broom, picks, rags, oil pot, roller, and diesel fuel for cleaning tools. If the area is very small, you can eliminate the roller and oil pot. The oil can be spread with a 5-gallon bucket and a brush. A small gas operated plate tamper can be used in place of the roller.

A 1 or 2 ton vibrating roller is excellent for small areas when a larger roller isn't specified. It's very hard to get a good smooth and level surface with a plate tamper.

Tamping the Edges

If edges along curbs or walls can't be rolled with the roller, they must be tamped with a plate tamper or a hand tamper before they have a chance to cool. Use a can of water with a spray attachment to moisten the asphalt ahead of the tamper. This will prevent the asphalt from sticking to the tamper plate. Most plate tampers are equipped with a small tank for diesel oil or water. The fluid drips onto the plate from a tube along the front, and the vibrating action sprays it onto the asphalt. Water will keep asphalt from sticking to the plate temper. Diesel oil must be used on rakes, shovels, and hand tampers. A bucket or spray can of diesel oil should be available at all times when paving, regardless of whether you're paving by hand or machine.

Paving Trenches

For trench paving, two lifts are advised but not essential unless more than 4 inches are needed. This is the best procedure when paving a trench if 4 inches of asphalt are needed: Dump enough mix in the trench to fill it to the top of the existing asphalt on each side. Tamp it well with a plate tamper or narrow roller that fits the trench width. When it's rolled, it should be compressed approximately 1 inch below the existing asphalt on each side. Spray the existing edges with an asphalt emulsion and dump in more asphalt. This time rake the edges so that only a small amount of fine material extends past the trench edge, and leave the asphalt 1/2 inch higher than the trench edges. Finally, roll this level with the existing surface.

If a large amount of trench paving is to be done, a spreader box can be modified so that it can be used to pave the trench. Plates can be spot welded in the box to narrow the opening to match the trench width. If using a spreader box, apply a tack coat to the edges first. Pave the trench using one lift, followed by a 5 or 8 ton roller. If the paving is to be more than 4 inches thick, two lifts may be desirable, regardless of whether a spreader box is used.

Paving Airport Runways

On airport runways where 95% or more compaction is required, you should have a thin layer density gauge. Check the density periodically. If the density is achieved, the rollers can move ahead. This eliminates wasting time rolling an area that has already reached 95%. If it's below 95%, you can reroll the area before it cools.

Estimating the Asphalt

To estimate the amount of asphalt needed, use this method: Take the square footage to be paved, times the inches of asphalt to be placed, then divide by 160 to get the tons of asphalt you need. Here's the formula:

$$SF \times depth\ in\ inches \div 160 = tons\ of\ asphalt$$

For small areas, remember this rule of thumb: One ton of asphalt will cover 80 square feet 2 inches deep.

Trenching

As in most excavation work, trench excavation requires the right equipment. There are sizes and types of trenching equipment appropriate for any job you have. Your first step is to decide which type of equipment will work best. There are four basic types of trench excavating equipment: wheel trencher, bucket line trencher, backhoe, and a dragline.

Trench Excavation Equipment

Generally you would use a wheel trencher for depths 6 feet and under where there are no obstacles along the line. You might use a backhoe if the trench is extremely wide or the trench walls must be sloped. You would probably also choose a hoe if you have to excavate around any obstructions such as utility lines. On small jobs, a backhoe is better because of the cost of moving wheel trenchers to a site. Where a long reach is needed and if the material being excavated is loose or soft, a dragline is best. In that case, a backhoe

Crumbing shoe is close to trencher buckets
Figure 27-1

would not be able to find a firm footing close enough to the edge of the trench to reach the trench bottom.

Let's look at the types of trenching equipment one at a time. We'll start with the wheel trencher.

Wheel Trenchers

Wheel trenchers have higher production rates than any other type of trenching equipment. Their only drawback is that most wheel trenchers are limited to a depth of 6 or 7 feet and a width of 24 inches. Although larger wheel trenchers are made, they're not widely available.

When you use a wheel trencher, it should be equipped with a crumbing shoe to keep the loose dirt from being left at the trench bottom behind the wheel. See Figure 27-1. As the trencher digs, the shoe forces dirt back into the trencher buckets.

A wheel trencher leaves very little dirt along the top edges of the trench. This makes cleaning the edges easy for the pipe crew. But wheel trenchers leave a large plug of dirt to remove by hand or with a hoe when you're working close to manholes and utility lines. See Figure 27-2.

A wheel trencher is more maneuverable than a bucket line trencher. It's excellent for short runs like shallow sewer services. But you need a good, level surface when working with trenchers. A wheel trencher can go up and down hills, but the ground has to be fairly level across the trench, from left to right at right angles to the trench. If it's not, the trencher will lean to one side or another. Leaning will cause the trench walls to be slanted. Not only is this hard on the trencher, but a trench out of plumb is a dangerous place for the pipe crew to work.

Setting a string line with a wheel trencher— The wheel trencher is usually a good choice for water lines because the depth is usually less than 5 feet deep. The water lines are generally placed after all the rough grading has been completed. When you excavate for a water pipe, set a string line to keep the trench in a straight line. A line is not usually necessary to give the correct depth. If only a minimum cover is needed for the water line, the trencher operator sets the grade from the level of the ground where the trencher works. In most cases the subgrade has already been cut. Depth calculation should be easy.

When a water line is being trenched through an area that hasn't been rough graded, you may need to use a grade line to set the depth. If this is the case, the surveyor will mark cut grades as well as stakes to set the trench line. You'll always need a grade string for line and grade when wheel trenchers are cutting sewer line or drain line trenches.

Bucket Line Trenchers

Bucket line trenchers are available for various depths and widths. You can vary the widths on most bucket line trenchers. If the trencher you have available is capable of trenching the depth you need but isn't set up for the width you want, bolt extension brackets to

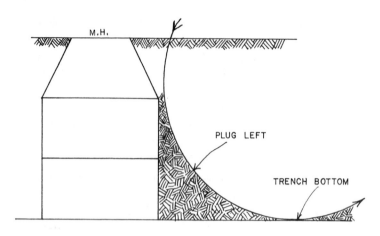

A   Wheel trencher

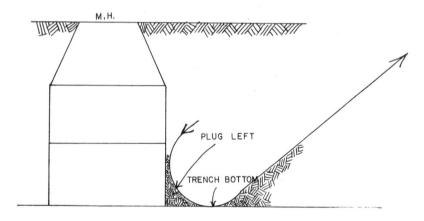

B   Bucket line trencher

Wheel trenchers leave a large plug of dirt
Figure 27-2

Two ripper teeth welded on top of trencher bucket
Figure 27-3

the sides of every other bucket. Weld one or two tooth sets to the brackets and set teeth in them. This is a simple job and shouldn't strain the bucket chain. A 2-foot wide bucket equipped with extensions can dig a 3-foot trench.

What if you need a 4-foot wide trench and you only have 2-foot buckets available? You can make brackets that attach to the bucket chain rather than to the buckets. When these brackets are attached for extra width, the trencher operator will usually need to use a lower gear on the crawl speed and a higher gear for the bucket line and conveyor speed. That compensates for the extra dirt being removed.

When production is dropping because you're trenching in hard-pan or very hard soil, attach ripper teeth to the top of the buckets. See Figure 27-3. Weld these teeth on every other bucket. They can

Skirts push excess dirt back into trench
Figure 27-4

be staggered so that two rippers are on one bucket and one ripper is on the next bucket. By setting rippers above the buckets, you can rip the ground ahead of the bucket teeth and improve production.

A bucket line trencher will leave a great deal of loose material along the top edge of the trench, especially at shallow depths. Skirts can be made to drag alongside the bucket line and push the material back in the trench. Figure 27-4 shows the skirts on a trencher. A man in the trench behind the bucket line should throw the loose material from the bottom of the trench back into the buckets. See Figure 27-5.

Trenches over 5 feet deep must be shored. The shoring should be placed close behind the digging operation so the man crumbing behind the trencher is always protected from a cave-in. During any trenching operation, watch the ground closely. If you notice subsidence or any cracks, stop trenching immediately until shoring

Crumbing and grading behind bucket line trencher
Figure 27-5

can be placed. Cracks may appear from 1 to 10 feet from the trench edge. When this happens, the supervisor must decide whether to continue with a trencher or use a backhoe so the trench walls can be sloped.

A bucket line trencher can do a good job of trenching close to a manhole, as shown in Figure 27-2. Because it has a smaller digging radius, it can move in closer to obstacles than a wheel trencher. A bucket line trencher also leaves a much smaller plug of soil next to obstructions, as shown in Figure 27-6. A backhoe is more efficient around utility lines.

Jumping utility lines with a bucket line— If you are trenching with a bucket line and there are several utility lines to jump, locate them by hand digging. Then use a shovel to dig about 6 inches in front of each line toward the trenching operation and level with the top

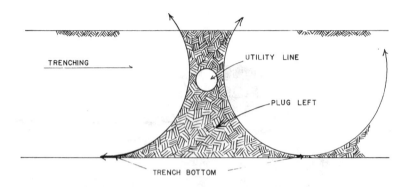

A   Utility plug left by wheel trencher

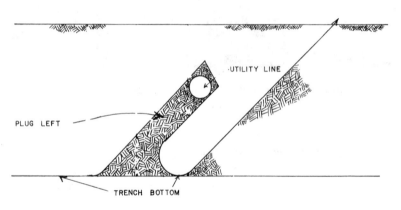

B   Utility plug left by bucket line trencher

Utility plugs left by trenchers
Figure 27-6

of the utility line. Place shredded newspaper or a few shovels full of white lime in the hole on the trencher's side. Then mark the spot by painting or staking away from the centerline so the trencher operator will see it. Finally, refill the hole.

As the bucket line nears the utility line, have someone watch the soil coming out of the trench. The observer should be ready to call to the operator when lime or newspaper shreds are pulled up. When that happens, the operator knows that the bucket line is only 6 inches away from the utility line. The operator should raise the stinger from the trench and pull ahead to just beyond the utility line. Move any loose dirt over the utility line so a visual check can be made.

Once the top of the bucket line is below the utility line, the trencher operator can dig with the trencher moving toward the utility line. The operator lowers the bucket line until it reaches the desired depth and trenches to within 6 inches of the line from the back side. See section B in Figure 27-6. Jumping utility lines in this way will leave the smallest possible plug. That means you'll have less tunneling to do, either by hand or with a hoe.

You can also jump utility lines by setting the trencher in behind the utility line and working away from the line. When the full depth has been reached, the trencher continues up the trench line. This leaves a larger plug of dirt but saves time for the trencher. A rubber-tired backhoe straddling the trench removes the plug the trencher left. This is possible only if the ground is firm enough to hold the weight of the hoe. The hoe can be kept occupied between utility plug work by supplying grading material, digging manholes, backfilling trenches, and setting pipe or manhole barrels.

Backhoes

A backhoe can be used any place a trencher can be used. A backhoe doesn't trench as rapidly but is excellent for short trenches and jumping utilities. There are many sizes and models of backhoes. Most have bucket sizes for nearly any job. Special round bottom, V-shaped, rock and trash buckets are also available. A small loader-backhoe combination may be the only equipment needed on a small trenching project. The backhoe can trench, supply bedding material, set pipe, and backfill. A backhoe is also a good piece of equipment for digging and setting manholes.

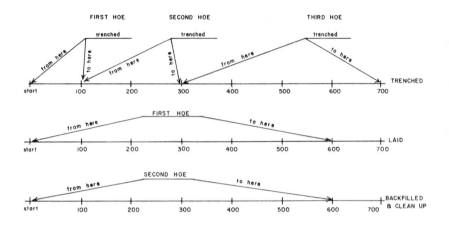

Three-hoe trenching pattern
Figure 27-7

Using a two- or three-hoe trenching pattern— Where utility cross-ings are a problem, you can get good production using two or possibly three backhoes. Here's how to do it. Assume a hoe can trench 100 feet every two hours on a particular job. You're going to use three hoes that also have front loader buckets. And assume the pipe crew is capable of grading and laying 100 feet of pipe an hour when a hoe sets the pipe. Set up a trenching pattern like this: The first hoe trenches for two hours and then begins setting pipe for the pipe crew. The second hoe, beginning 100 feet ahead of the first hoe, trenches for four hours and then begins backfilling the 200 feet of pipe that have by now been laid by the first hoe. The third hoe, beginning 300 feet ahead of the first hoe, would con-tinue to trench the entire eight hours.

Using this plan, by the end of the day you have trenched 700 feet of ground, laid 600 feet of pipe, and it's all backfilled. See Figure 27-7. The next morning there would be 100 feet left for the pipe crew to grade when the trenching pattern begins again. In this way, the pipe laying is kept within 100 feet of the hoe trenching. See

Two hoes in a three-hoe trenching pattern
Figure 27-8

Figure 27-8. The number of men used for laying pipe depends on the number of hoes and the amount of footage being trenched.

This "leap frog" method of trenching works very well in firm soil. The backhoe needs to be able to straddle the trench while digging, yet move off the trench without caving in the sides. Straddling a trench with a small backhoe is very simple for an experienced operator.

If the trench sides are too soft for shoring, you can still use a backhoe. Instead of vertical trench walls, the trench walls must be sloped a minimum of 3/4:1. For every 1 foot the trench rises vertically, it must slope out 3/4 of a foot horizontally. There are a couple of disadvantages, though. Of course, this increases the amount of dirt the hoe must move and cuts the production considerably.

Sloping the trench sides also makes it harder to use a three-hoe trenching team. When the first hoe gets to the point where the second hoe started, the first hoe can't straddle a trench that's being sloped. The first hoe operator must move off the trench. He can then tie his trench into the trench started by the second hoe by digging from the side. But it's hard to keep the spoil dirt away from the trench edge when you dig at an angle. Usually the hoe operator must stop trenching and move the spoil pile and then begin again.

If your progress is slowed by very hard ground, you can attach rippers to the hoe's bucket. Many buckets are designed with slots for ripper shanks. If not, you can easily attach rippers by welding brackets on the bucket.

Trenching around utility lines with a hoe— When trenching around water mains, have a drainage plan worked out in your mind in case a main is broken. If this happens, water has to be channeled or pumped away from the work area. Small copper or plastic water lines can be closed to stop their flow. Copper service lines can be bent and hammered shut. Plastic lines can be crimped with vice-grip pliers. If galvanized pipe is cut, try to drive a redwood plug into the pipe to stop the leak. Shut the water off first if you know the location of the shut-off valve.

Always mark utility lines well in advance so that production isn't interrupted by a broken line. In a very short time a broken water main can cause major damage and flood a wide area, including adjacent buildings.

If a trench crosses utility lines at other than a right angle, use a backhoe to keep hand tunneling to a minimum. By digging first on one side of the trench and then the other, you can work parallel to the utility lines. This is true even if there is more than one line to jump in the same area. Figure 27-9 shows how easily a backhoe can work around utility lines at a 45-degree angle, eliminating the need for hand tunneling.

Draglines

A dragline is needed when the trench is deep and the soil is soft. Draglines are excellent for cleaning irrigation ditches or channels

**A backhoe trenches easily between utility lines
Figure 27-9**

Trencher with portable conveyor
Figure 27-10

where the reach required is too great for a backhoe. The major advantage of a dragline over a hoe is its greater reach. If an area you're excavating can't be reached by a backhoe, you're probably going to have to use a dragline. But a dragline isn't as efficient as a hoe. The hydraulically-operated hoe has a much faster cycle time than the cable-operated dragline.

You can use a dragline with an open bucket similar to a hoe bucket, or a clam shell. A few other buckets are available for special projects.

Planning a trenching job can be both a creative and a rewarding task. Many jobs present opportunities for innovative solutions. In Figure 27-10 the crew is digging a very deep trench with limited room for a spoil pile. A conveyor belt was set up so the trencher's conveyor dumps into a portable conveyor. This dumps the dirt back into the trench behind the pipe laying operation. In this way the trencher can get rid of the spoil and backfill at the same time. Always look for better, faster, and more professional ways to get the job done.

Trench Shoring

The introduction of hydraulic shoring to the construction industry several years ago has made working in a trench safer, and a lot easier. Hydraulic shoring is much faster than any other bracing system. It's also safer than manually-operated screw jacks because no one has to work in the trench while placing the shoring. You can place the entire unit from the top of the trench wall.

Various lengths and widths of hydraulic shoring are available. You can also use more than one length of shoring on each point along the trench. For example, if only 6-foot lengths are available and the trench is 12 feet deep, you can use two 6-foot lengths to make a 12-foot shore. That's one of the advantages of hydraulic shoring. The first 6-foot shore is jacked into place at the top of the trench. Then, using a long release tool, the second shore is placed under the first. Setting hydraulic shoring 6 feet long is fast because each shore is light and easy to handle.

Every shoring section has at least two pump cylinders. See Figure 28-1. Longer lengths have more cylinders. If the shoring is

Hydraulic shore, pump tank and release tool
Figure 28-1

longer than 7 feet, two men will be needed to handle the shore weight. See Figure 28-2.

Even though the shoring uses hydraulic principles, the fluid used in the jacks isn't hydraulic oil. There are several special fluids made for hydraulic shoring. They're much cleaner and less expensive to use than oil.

Setting Hydraulic Shoring

Use the following procedure when placing hydraulic shoring. The shoring should be set on the ground near the trench in a folded or collapsed position. Locate the hydraulic quick coupler at the base of one of the jacks. Turn the jack cylinder so that the quick coupler is facing upward. Lift the top aluminum plank so the shore is open or unfolded. Once the shore is open, hook the pump hose to the quick coupler which is now readily accessible. Notice in

Two men placing a long hydraulic jack
Figure 28-2

Figure 28-1 that another hose runs from the top jack to the bottom jack so that both cylinders can be activated by the same quick coupler.

The hydraulic pump tank has a wing nut valve on top. This valve must be open to connect the quick coupler. After the quick coupler is connected, push down on the top aluminum plank so the hydraulic cylinders are completely compressed. With the wing valve open, the fluid in the cylinders flows from the jacks to the tank. Now the hydraulic shoring is ready for placing in the trench.

Fold or collapse the shoring again and ease it into the trench with the jack hook (the "release tool") attached to the shoring handle. The hose, hook, tank and quick coupler should all be on the same side of the trench. Hold the handle on the plank that will be on the far side of the trench. That way the shore won't spring open until it's roughly at the right level in the trench.

Look at Figure 28-2 again. Once you have the shoring where you want it, release the far plank handle and hold the near side plank by hooking the jack hook to the near side handle. This allows the shoring to spring open. Now steady the shoring with one hand, holding it in place with the hook. With your hand, twist the wing valve shut. Pump the hydraulic tank arm until the jacks spread enough to push the aluminum planks against the walls of the trench. You should feel a firm back pressure on the tank handle. Now, while still holding pressure on the pump handle, use the release tool to pop the hose coupler free from the jack quick coupler. Most tanks have a pressure gauge to indicate when the pressure is high enough to release the coupler. This completes the installation. Follow this procedure whenever setting hydraulic shoring.

Removing Hydraulic Shoring

To remove hydraulic shoring, stick the release tool through the handle on the quick coupler side of the shoring. With the hook facing the center of the trench and the cup at the end of the release tool turned inward, place the cup over the quick coupler. Exert pressure against it by pushing out against the handle with the release tool. The fluid will spray out, releasing the pressure on the aluminum planks.

Keep putting pressure against the quick coupler until you have a large opening between the shoring and the top of the trench. Then pull the release tool up with a jerk so the hook catches the handle on the shore plank. You can now pull the shore from the trench. It will collapse as you pull the handle of the plank on the far side of the trench. If the shoring is long, pull it far enough with the hook to wedge the shoring on an angle against the opposite side of the trench. Then remove the hook (release tool) and pull the shoring the remainder of the way by hand.

On some jobs you may use hydraulic shoring that's too long and heavy for one man to lift. You'll need a second man to help. If the trench is extremely wide or deep, you may need a hook and rope to pull the far side handle so the shoring will collapse.

Placement of Shoring

When you set hydraulic shoring, try to place the shore planks where the trench wall is straight and smooth. If the trench wall is rough or if there's a void behind the jack, pressurizing the shore may bend the aluminum plank. If you can't find a smooth area, place blocks behind the jack to prevent bending.

When you work in gravel or sandy soil, too much hydraulic pressure will make the trench wall crumble. In soil like this, pump just enough to create a firm pressure against the walls. You must use sheeting if the ground is too loose. Slide the sheeting into the trench and then hold it in place with hydraulic shoring.

Check local and federal safety regulations on trench shoring before you do any shoring. These regulations will spell out very clearly how far the shoring must be set apart and how thick the sheeting must be at various trench depths.

Shields

You may work in areas where ground conditions are so bad that shoring won't hold the trench walls, even when backed with sheeting. If your space is limited and it's not possible to slope the bank back from the trench bottom, a shield must be used.

There are many types and brands of shields to choose from. Some have adjustable widths and some have fixed widths. A simple shield could consist of two sheets of steel or lighter metal separated by welded steel braces. An open area in the center between the braces allows a pipe to be lowered through the shield to the trench bottom. The shields must be pulled along by the hoe as the pipe is laid. See Figure 28-3.

Some more elaborate shields have a hopper at the front that feeds gravel to the bottom of the shield and a screed that spreads the gravel to the correct grade. As the shield moves forward, pipe is lowered through the center opening and a hopper at the back of the shield backfills gravel over the pipe. The excavated material is backfilled right up to the back of the shield. A hydraulic ram pushing against the backfill propels the shield.

Aluminum shield pulled by backhoe
Figure 28-3

Equipment like this is usually used by contractors who specialize in large underground projects. But whether you're working on large projects or small ones, it's essential that you shore any vertical trench wall over 5 feet deep. *Do not place men in a trench deeper than 5 feet without some type of approved shoring,* regardless of how solid the ground appears to be.

Manhole Shoring

It's only slightly more difficult to shore a manhole than a trench, even when the manhole must be free of braces so you can raise or lower materials through it. You can do this by setting a steel tube into the hole. These tubes or shields have slots cut into the sides to allow them to set over sewer pipes. They also have clips at the top so a second shield can be placed on top of the first and hooked into place. See Figure 28-4.

Manhole shield with clips at top
Figure 28-4

If the manhole is deep and overdug, it's a good idea to center the tube. Look at Figure 28-5. Center the tube by placing hydraulic shoring between the manhole wall and the tube. Centering and bracing the tube will prevent the worker's legs from being trapped between the pipe and the shield in case of a cave-in.

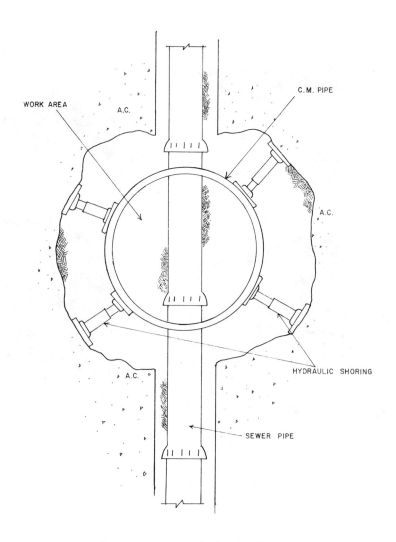

Open center manhole shoring
Figure 28-5

29

Laying
Water Pipe

This chapter will explain the essential principles of laying water pipe, and show you some of the practical tips used by professionals in the field. Most companies that manufacture water pipe offer excellent handbooks that show how to lay the particular pipe they sell. The sales representative who supplies the pipe can get the copies you need.

Laying the Water Main

Good installation practice begins with unloading the pipe. Spread the pipe along the trench so that you have the required footage between each station. Some pipe, like 6 and 8 inch plastic or asbestos cement pipe, can be laid by hand. All cast iron water line, because of its weight, must be set with a crane. The best crane for laying water main is a truck-mounted crane. Load the pallets of pipe on the truck and then, using the crane, unload them directly from the truck to the trench as shown in Figure 29-1.

Truck-mounted crane laying pipe
Figure 29-1

One main advantage in laying water pipe, as opposed to sewer or drain line, is that the trench bottom doesn't need to be fine graded. Water pipe is laid on two mounds of dirt about 36 inches back from each joint. See Figure 29-2.

Your biggest concern in laying water main is keeping a constant elevation and line. Each pipe length must line up with the next. If you use flanged pipe, extra effort is required to get the bolt holes lined up. You can make slip ring joints on a slight angle. However, if the slip ring joint is pushed together at too great an angle, the sliding ring won't seal properly. Forcing the ring may cause the sealing ring to roll out of its seat. This will result in a high volume leak when the water is turned on. The various types of joints are shown in Figure 29-3.

It usually takes three men to lay water main. One man hooks the pipe into place and two do the actual laying. When you use a slip

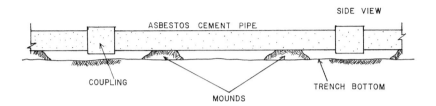

ASBESTOS CEMENT PIPE

SIDE VIEW

COUPLING

MOUNDS

TRENCH BOTTOM

Water pipe is laid on mounds of dirt
Figure 29-2

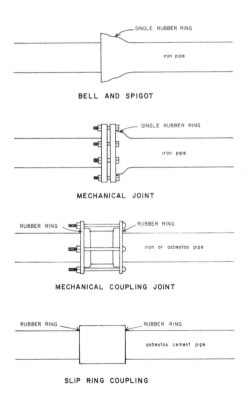

SINGLE RUBBER RING

iron pipe

BELL AND SPIGOT

SINGLE RUBBER RING

iron pipe

MECHANICAL JOINT

RUBBER RING

RUBBER RING

iron or asbestos pipe

MECHANICAL COUPLING JOINT

RUBBER RING

RUBBER RING

asbestos cement pipe

SLIP RING COUPLING

Pipe joints
Figure 29-3

ring pipe, always lubricate the spigot end and not the bell end of the joint. Both ends may be lubricated, but never connect pipe with the spigot end unlubricated.

The person helping the pipe layers does the lubricating and guides the pipe end into place. A rubber glove is handy to use when applying the lubricant. Cut the glove off just above the thumb so the hand will slide in and out of the glove without using both hands.

The pipe layer needs a bar to move the pipe into place. Putting a piece of wood between the bar and the pipe protects the pipe when pushing the pipe joints together.

In most cases water line trench is from 3 to 5 feet deep. The pipe layers and their helper can use the material from the spoil pile along the trench to build the mounds on which the pipe rests. If they can't reach the material, a fourth person is needed to shovel dirt for the mounds.

Using Transite Pipe

Laying the main using slip ring joints is a simple operation. See Figure 29-4. A truck-mounted crane is holding the cable steady while the pipe layer's helper lubricates the end of the 6-inch water line pipe. A quick-release pipe clamp holds the pipe, so no sling is needed.

It's more involved and time consuming to place valves, tees, crosses, bends, and blow offs, and to make water service taps. Slip ring fittings are usually used with asbestos, plastic, or iron pipe for water mains. Most asbestos cement water line (also called transite line) comes in 13-foot lengths. Shorter lengths with milled ends are also available in lengths of 6 feet 6 inches and 3 feet 3 inches. You can use the shorter lengths in combination with the full lengths to avoid having to cut pipe when you approach a valve or fitting.

MOA and MEE pipe— Short lengths of pipe that are milled the entire length rather than just on the two ends are also available. When a pipe is milled end to end, it's referred to as a *MOA* pipe.

Pipe layer's helper lubricating joint
Figure 29-4

MOA stands for *machined over all*. The short lengths that are only milled on each end are referred to as *MEE* sections, *machined each end*. See Figure 29-5.

The advantage of the MOA section is that it can be cut to any length and used without milling the cut end. But remember to bevel the cut end with a rough file so that it slides smoothly past the rubber ring. See Figure 29-6. Use a MOA section of pipe when you tie into another pipe or into a fitting that is stationary. This is because the two couplings can be pushed easily from the MOA to

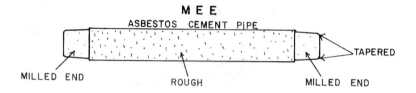

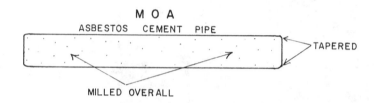

MEE and MOA pipe sections
Figure 29-5

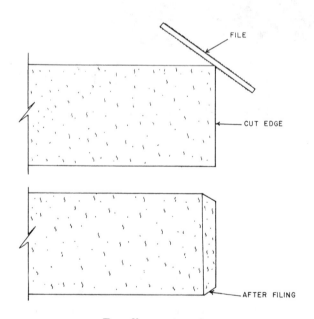

Beveling cut end
Figure 29-6

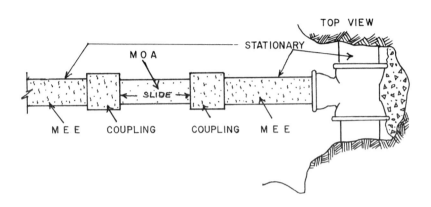

MOA section placed between two stationary pipes
Figure 29-7

the MEE pipe. Once the MOA pipe has been placed between the two stationary pipes, the couplings are pushed in opposite directions to make the connection. See Figure 29-7.

Cutting transite pipe— Never use a compression type cutter on plastic or asbestos cement pipe. Use a cutter with a knife-like rotating blade or blades on plastic or transite pipe. The cutter cuts slightly deeper with each turn until the pipe is finally severed. On transit, use a portable or truck-mounted milling tool to mill and bevel the end of MEE pipe once it has been cut. Hand operated milling and cutting tools work as well as power cutters and millers, but it takes longer.

Plastic pipe doesn't need milling, just end beveling. Because plastic pipe is smooth all over, you can use any piece cut to any length in the same way as you would a MOA transite pipe.

Locating transite pipe— Job specifications commonly require that a bare copper wire be laid alongside or over asbestos cement or plastic pipe so that its precise location can be determined later by a pipe locator. Usually two insulated copper wires are required at each valve riser so the wires won't touch when the metal detector is attached. If they touch, it causes a grounding effect, and the metal detector won't work properly. The coated wires should be on the outside of the valve riser. Without the copper wire, a metal detector can't locate an asbestos line.

Kicker blocks— Your job specifications may also require that valves and fittings be set on concrete or redwood blocks. All specifications for laying asbestos cement, plastic, and cast iron water line will require that a kicker block be poured at each bend or angle. Figure 29-8 shows various locations where kicker blocks are needed.

Asbestos cement or plastic pipe can be tied into cast iron or steel pipe by using fittings designed for this purpose. If specifications allow, consider using plastic pipe with the strength of cast iron as an alternative to iron pipe. Plastic pipe is much easier to handle because it's lighter. Plastic and iron pipe come in various lengths from 13 to 20 feet.

Water Service Taps

There are a variety of ways to make water service taps to asbestos cement, plastic, and iron water lines. Some asbestos couplings have built-in service taps. A service clamp is the most common connection. It can be strapped right to the main. Then you can drill and thread a hole in the pipe for a corporation stop. Always drill the tap hole after the clamp is on the main. A typical service strap hookup is shown in Figure 29-9A.

Any type of service pipe can be connected to the corporation stop. The service line may be plastic, galvanized, or copper, depending on which is specified. You can screw galvanized pipe directly to the corporation stop with a coupling. Unless glue or sweat fittings are used, flare plastic and copper pipe to make a connection. Slide the coupling nut from the corporation stop over the plastic or copper pipe before you flare the end. Once you've flared the end with a flaring tool, attach the service line by sliding the coupling nut down the pipe and screwing it into the corporation stop.

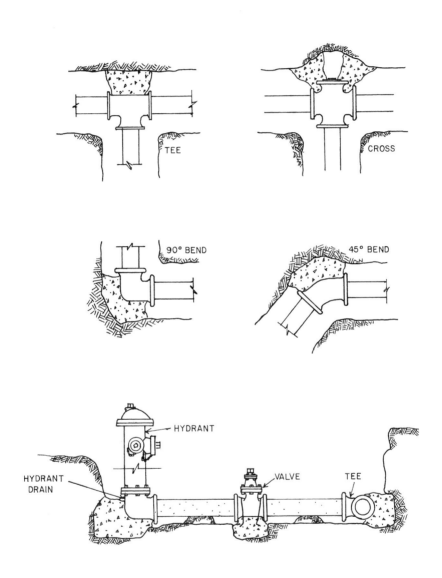

Concrete kicker blocks
Figure 29-8

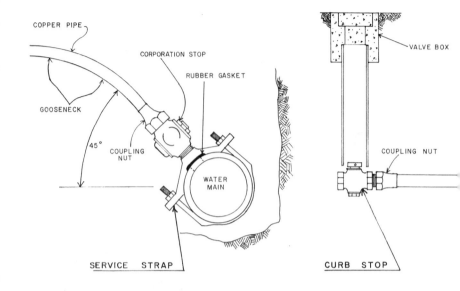

A   Water service tap using service strap   **B   Water service curb stop**

Typical water service hookups
Figure 29-9

When you order corporation stops, specify the type of threads you need to connect to. All water service lines should have a valve, curb stop, or meter with a valve box and riser at the property line. See Figure 29-9B.

If you use copper for the service lines, all joints must be soldered with a copper coupling. Before you solder copper fittings together, be sure all surfaces are clean. Use a fine emery cloth to sand any area that will be soldered. Then apply the acid solution and solder. Plastic fittings must also be clean and free of moisture before you apply the plastic glue. Keep all galvanized fittings clean and use a liberal amount of joint compound or joint tape at each joint.

Attatching a Service Clamp

When you're using a service clamp for a water service tap, use this procedure. Attach the service clamp firmly to the main, screwing the bolts on each side down evenly until it is tight (55-65 foot-pounds with a torsion wrench). Unless the main is very shallow, set the service clamp at a 45-degree angle to the main. After the service clamp is secure, drill the service tap hole and screw the corporation stop onto the service clamp. Asbestos cement and plastic pipe can be drilled easily with a masonry drill bit. Once the hole is drilled, attach the service line.

The service line should have a gooseneck in it as it leaves the main, as shown in Figure 29-9A. The gooseneck will absorb any movement or settlement without damaging the service connection in any way. The maximum size of the service tap is limited and varies with the size of the pipe being tapped. Service clamp taps range in size from 1/2 inch to 3 inches in diameter. Check with your supplier to get the tap limitations.

Tapping a Live Line

When tapping a steel or cast iron main or a live line, use a tapping drill rather than the standard drill and masonry drill bit that you use for dry asbestos cement pipe. Some tapping tools screw to the corporation stop and others are secured directly to the main.

Here's how to tap through a corporation stop on a main that's under pressure. Attach the tapping tool to the corporation stop, making sure the corporation stop is open. The tapping tool should have a ratchet handle to turn the drill bit. The bit must be ratcheted slowly through the wall of the water main. You should feel some tension release on the ratchet and there may be some water leakage from the tapping tool when the main is pierced. Be sure the drill bit goes completely through the wall of the main. You may get some leakage as soon as the bit breaks through the wall, but this doesn't mean that the tap bit has made a hole the full diameter needed for the service tap.

Most tapping tools have a long drill bit. Be careful not to drill completely through the pipe, especially on small diameter mains. Once the tap has been made, back the drill bit out of the main and the corporation stop. Close the stop before detaching the tapping tool. Be sure the service lines have been attached and completed and the corporation stops are open before backfilling.

Backfilling

Pour all bends and fittings that require thrust or kicker blocks before backfilling. The job specifications will give you the required size of the thrust blocks. Figure 29-8 shows the correct positioning of thrust blocks and where you should pour them.

Once the system has been completed, begin the backfill. The initial backfill must be done carefully because the pipe is sitting on mounds. Any sudden jar from backfill material at the joint may fracture the pipe. Carefully tamp backfill material under the pipe to fill any voids. For asbestos cement pipe, sand is an excellent initial backfill material. Be very careful while backfilling. Don't let any large chunks of dirt or rocks lodge against the pipe.

Laying water mains takes a great deal of care. Leaks are hard to locate, and when found, often troublesome to repair. Usually the repairs must be made in tight and muddy quarters.

Pressure Testing the Main

In most cases, you can begin testing after the trench has been backfilled and compacted or water jetted. But check the job specifications for test requirements. Some agencies require that all joints and fittings be exposed during a pressure test, and that just enough backfill be placed on the pipe to keep it from moving under pressure. In this case, backfilling and jetting will be completed after the pressure tests. Your job specifications will also specify the length of time you must allow concrete kicker blocks to cure before applying pressure. It may be as long as 24 hours.

In some cases the pressure test consists simply of slowly opening the valve from the live main until the new line is filled. Then, after a few hours the inspector makes a visual check of all the joints and fittings.

A more difficult test requires that you pump extra pressure into the line. Complete backfilling and compacting is usually required before this type of test begins. Start by slowly opening the valve from the live main to the new main. Crack open the hydrants along the system to release air in the line. You'll need to shut them quickly when all the air has escaped and the water begins to flow evenly. Drain off just enough water to release the air but not enough to lose any chlorination that may be in the line.

Once the air is released and the new main is full, shut down the valve feeding the line. You'll need a pump to add the required pressure. Usually the pressure specified will be around 150 to 200 pounds per square inch. The pump draws water from a 55-gallon drum and pumps this water through a discharge hose attached to a hydrant in the system. There should be a valve and a pressure gauge before the coupling attaching the hose to the hydrant. Close the valve when the required pressure is reached. and shut the pump down.

The amount of time the test pressure must be held varies with different agencies. Some agencies test by the pressure lost. Most agencies measure the gallons needed to fill the line after a period of time. This is done by measuring the amount of water drawn from the 55-gallon drum when the pump refills the line to the original pressure. The amount of leakage allowed varies from one agency to another. The specifications will give you the allowable amount.

Chlorinating the Main

Once the results of the pressure test are acceptable, the main must be chlorinated. You can use one of two methods. In the first, place a specified number of calcium hypochlorite tablets in each joint of pipe as the pipe is laid. Fasten the tablets to the inside of the pipe with permitex or rubber cement purchased at any auto supply store or pipe supply company. Apply permitex or cement to only one side of the tablet, leaving the remainder of the tablet exposed to the water. Your job specifications will dictate the exact number of tablets needed in each pipe joint.

If you place the tablets while laying the line, the line can't be flushed before testing. You must take great care to keep dirt and foreign matter out of the pipe when laying it if you can't flush the line.

Chlorination testing is done after the pressure test, when the main is already full of water and the tablets have dissolved. Open the valve on the live main slightly to allow water into the new main. At the same time a hydrant at the extreme end of the new system should be opened slightly until water runs out slowly. After several gallons have run through, the water should be fairly clear. At this point, the agency will test a water sample. If the hypochlorite tablets have dissolved correctly and care was taken to keep dirt out of the line, you'll pass the chlorination test easily. The main must remain full of water for a certain time before the test is made, but the time needed for the pressure test usually meets this requirement.

If you don't put tablets in the pipe when you lay them, calcium hypochlorite must be pumped into the system. This is the second method of chlorinating the system. With the main full of water and flushed, turn the controlling valve off. Connect a hydrant or service valve close to the controlling valve at the live main end of the line to a pump and reservoir of calcium hypochlorite dissolved in water. The reservoir for the mixture of water and calcium hypochlorite will usually have to be a water truck, because you need enough volume to fill the line. Open a hydrant or service valve at the far end of the main. Then pump the liquid calcium hypochlorite into the main until it starts coming from the hydrant valve at the far end. You'll be able to smell the calcium hypochlorite as it comes out of the hydrant.

Once this occurs, shut everything off. The line is now chlorinated and must sit a certain length of time before a sample can be taken. Once you've passed the chlorination test, flush the line well for several hours.

Once the line is well-flushed, it's ready for the bacterial test. If it doesn't pass the bacterial test, flush the main for a longer period, and retest. Some water districts may require the pressure and chlorination test in reverse order. Check your job specifications.

Using a Good Tie-out System

If water valves are covered before risers and valve boxes are placed, make sure you use a good tie-out system so they can be located later. When possible, place riser pipes over valves and let them extend above the existing ground about 2 feet. After landscaping is finished, cut them off enough so a valve box can be placed over the riser. If you leave the risers 2 feet high, they won't get filled with dirt while grading is being done.

If valves are in a road where grading and paving must be done, the riser pipe should be left enough below road subgrade so they aren't ripped up during grading and compaction. Cap or cover the risers to keep them from being filled with dirt.

After the road has been paved, locate the valves, extend the risers, and place the valve boxes. If the water line is shallow, it may be easier to leave the risers off completely until after the road has been paved. Either way, you need a good tie-out system to locate the valves after the paving of the street is completed.

Remember, every water district, private or public, requires certain brands and types of hydrants, valves, fittings, and pipe be used. This is why it's so important to check the plans and job specifications carefully before beginning work.

Laying
Sewer Pipe

The starting point for every sewer laying job is unloading the pipe correctly. Where you unload the pipe depends on whether the pipe will be placed by equipment or by hand. If you use equipment with a swing boom, you have two choices. You can place the pipe along the edge of the trench, or place it far enough from the trench edge so the equipment can drive between the trench and the pipe. When you use fixed boom equipment that will raise and lower a load but doesn't swing, unload the pipe along the trench edge. You should also place the pipe along the trench edge, as in Figure 30-1, if it's going to be lowered by hand or rope. However, never try to save money by hand-lifting long or heavy sewer pipe.

If pipe is being unloaded on pallets, it's important to know how many linear feet are on each pallet. Space the pallets so you have the right length of pipe between each trench station, as shown in Figure 30-2. Your pipe laying operation will be delayed if the crew and equipment must go up or down the line for additional pipe.

A well-prepared set-up for laying pipe
Figure 30-1

Pallets of pipe spaced between trench stations
Figure 30-2

Crane with nylon sling unloads pipe
Figure 30-3

Be prepared for the pipe when it arrives. Have an unloading pattern already worked out and your equipment available. To be sure that you have the correct equipment for the job, you must know the weight of the pipe or pallets. You also need to have the right slings or hooks available to unload the pipe or pallets without damage. See Figure 30-3.

The manufacturer can supply you with the weight information and explain how the pipe will be loaded. He can also suggest how to unload the pipe. Many pipe manufacturers will supply the slings or hooks needed for unloading. Talk with the manufacturer before you receive the shipment. A few manufacturers will even do the unloading, but you still must direct the man unloading the pipe to the correct area.

Preparing and Grading the Trench

When the pipe is unloaded and any required crushed rock or sand is dumped in convenient locations, you can begin preparations for laying the pipe. Dig the trench, then shore it and set a grade line. Refer to the chapters on trench excavating, shoring trenches, and grade line setting for detailed information on how to do this. When the shoring and grade line are placed, begin the hand grading. If no special bedding material is specified, you can use the trench spoil.

An improperly graded trench can cause serious problems. It can make sliding the pipe together difficult, especially if you're laying large pipe. And poor grading can interrupt the flow through the pipe. A poorly compacted trench bottom can also cause the pipe to break from settlement after the trench has been backfilled and compacted or jetted. It's important to have an experienced crew doing the grading and laying.

You need two or three people for grading. One person in the trench does the actual grading while one or two are on top of the trench. One can shovel material down to the one who's grading as it's needed. Or, if you're using aggregate or sand, he can direct the loader where to place it. If you require a top grade line, or if you use a top-side laser, the third person holds the grade rod and checks the grade. If you use a bottom grade line or bottom laser, the one doing the grading can check his own grade. He can also do it himself if you use a top grade line and the trench is shallow enough so the grade mark or laser receiver can easily be seen.

It's important that the material you use as fill for the trench bottom has a good moisture content so it will compact well. Don't use dry and dusty material. It will settle when jetting the trench. In order to make the trench bottom as firm as possible, the worker in the trench should pack the material down with his feet before he makes the final trim with his shovel.

If you need to use a pneumatic plate tamper, a second grading crew should follow the tamper to make any fills needed for the final trim. If sand or aggregate is required, a hoe or loader should dump material to the person grading. The one on top of the trench can show the equipment operator how much to dump and where it's needed.

If you have a choice of the type of aggregate to use, request 1/2-inch crushed rock. You can also use fine washed rock or sand, but crushed rock is excellent. It makes grading easy, it doesn't settle and it compacts with little effort. When the trench bottom is wet or soft, you must use crushed rock to provide a firm bottom for the pipe. Cut the trench far enough below the final grade to allow for the extra rock needed for firmness. If the trench bottom is sponge-like under pressure from your foot, it's too soft for laying pipe.

Most inspectors require that each pipe be checked for grade with a string line. You can do this with a surveyor's level or a laser beam as well. Some inspectors will allow the pipe layer to use a carpenter's level to check each pipe if the inspector is confident the trench was graded properly. In this case, the pipe layer can check each pipe with a carpenter's level to be sure it has the correct flow. If you use a carpenter's level, set the first few pipes with the string line so the pipe layer is sure these are at the correct pitch. Check these with the carpenter's level so you know how much bubble tilt is needed. Then set the rest of the line using that same bubble tilt, with a periodic check using the string line.

Laying the Pipe

The person laying the pipe must have three tools: a square nose shovel, a bar, and a level. When you shove the pipe together using a bar, place a block of wood between the pipe and the bar to prevent the pipe from being chipped. The pipe layer's helper should have a supply of lubricant and an applicator. If rubber couplings are used, they'll need a wrench to tighten the couplings.

You only need two people in the trench for laying sewer pipe, with one exception. If you use screw-type rubber couplings, the pipe laying may go faster than the pipe layer's helper can tighten the screws. When this happens, a third person must follow to screw the couplings tight.

Three Common Sewer Pipe Couplings

Figure 30-4 shows three common types of sewer pipe couplings or joints. When you use bell and spigot couplings, the bell always faces upstream and the spigot is pushed into the bell end. If solid coupling pipes are used, place the coupling on the upstream end of the pipe in the ditch before the next pipe is laid. Or you can place the coupling on the pipe before either are laid in the trench. Solid coupling or rubber coupling pipe usually comes with the coupling already attached to the end. If you lay the first pipe with the coupling facing upstream, all couplings will be correct.

Before joining bell and spigot or solid coupling pipe together, brush the rubber end with a lubricant supplied by the pipe manufacturer. The bell end is usually the end that needs lubricating on clay bell and spigot pipe. If you're laying solid coupling asbestos pipe, the spigot end must be lubricated. In cold weather, lubricate both ends of the pipe to make joining easier.

The pipe layer needs to dig a bell hole for each pipe. Bell-end pipe needs a deeper bell hole than solid coupling pipe. See Figure 30-4. The pipe layer and helper must take great care in keeping the joints straight and free from dirt, sand, or gravel. If any foreign matter gets into the joint, it's likely to leak when tested.

If the rubber ring and spigot end of solid coupling pipe aren't well lubricated or they get dirty, the rubber sealing ring may be dislodged when the joint is shoved together. If you fail to line up the two ends this will also cause the rubber ring to roll out of place. The pipe may appear to be seated correctly, but the pipe layer can usually tell that the ring has rolled because he must apply a lot more pressure to get the pipe seated. If more pressure than normal is needed to push the two joints together, check the joints careful-

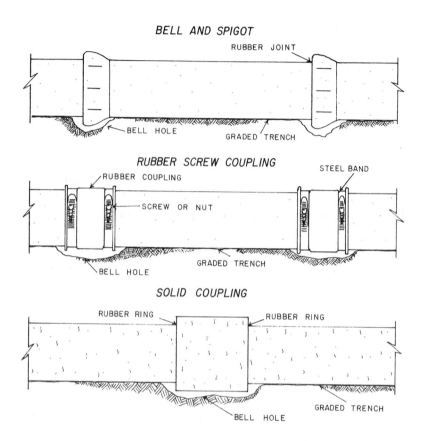

Three types of sewer pipe joints
Figure 30-4

ly. If the pipe hasn't seated properly the joints must be pulled apart and cleaned, relubricated, and the ring reset before they can be shoved together again.

It's easy to tell when the rubber coupling or bell and spigot type pipe are not seated correctly. The spigot won't seat fully in the bell. The rubber seals on bell and spigot clay sewer pipe are stationary and will not roll or move. If it doesn't seat fully, it's probably because the two pipe sections aren't aligned correctly. Also, there could be dirt in the bell end.

Laying Pipe at Manholes and Laterals
Every gravity flow sewer line should be laid from the downstream end of the line to the upstream end. When using smaller size pipe, lay them through manholes rather than leaving out a section of pipe at each manhole. Large pipe, depending on the type being used, may also be laid through the manhole. You can break the top out after the manhole bottom is poured and has set up around it. The tops of large pipe should be broken out before the manhole barrels are set. Use a jackhammer if needed.

Of course this doesn't apply if iron pipe is used. You won't lay iron pipe through the manhole because it's too much trouble to cut the top out.

If you're running any sewer laterals from the main, be sure to set the fittings, Y's or tees next to the trench where they should be placed. Look back to Figure 30-1. Proper arrangement of the fittings along the trench will prevent the pipe crew from laying straight pipe past the spot where a fitting is to be placed. The detailed drawing on your plans or specifications will show the angle the Y's and tees must take when laid.

The depth of the trench and the location of parallel pipe determine the angle of rise the service branches will have from the main line to the property line. See Figure 30-5. The specifications should tell you how many bends and how sharp a bend can be used on lateral runs. Job specs also show you where clean-outs are required. The main must be laid and the Y's and tees must be in place before you can begin any lateral trenching.

At the end of the day, plug all open pipe ends and place enough dirt or aggregate over the pipe to protect it. This helps keep vandals from dropping pipe or rocks into the trench and breaking newly laid pipe. And plugging the pipe ends keeps small animals from entering the pipe.

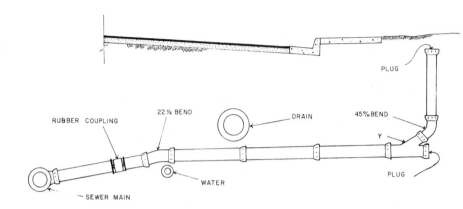

Sewer service branch
Figure 30-5

Repairing Broken Pipes

If a pipe or fitting in the middle of a line is broken, here's how to repair it. Expose and clean the main three joints on each side of the break. Remove all material on top of and on the sides of the main. Cut out the broken pipe and remove the pieces. You need five people for inserting a section, two on each side and one in the center with the new pipe. Lift the sewer line high enough to buckle the new section of pipe in place. This procedure works only with short lengths of small pipe that can easily be lifted by hand.

If the pipe is too large to lift by hand, the procedure is different. Assuming the pipe is in 6-foot lengths of 12-inch pipe, remove the broken section. Cut the bell off the good end of the pipe left in the trench, approximately 1 foot in back of the bell. Now cut 1 foot off the spigot end of the remaining pipe on the upstream side of the broken length. Take the new 6-foot replacement section of pipe and add the 1-foot section of bell and the 1-foot section of spigot end to it.

Slide two rubber couplings on the two cut ends of pipe. The couplings must be set far enough onto the two ends so that no rubber protrudes past the cut end of the pipe. If the rubber coupling has a rubber strip inside as a divider, cut it out so the whole coupling will slide onto the cut end. The coupling may need lubricating before it will slide fully over the pipe. Lower the pipe with the added 1-foot sections and couplings into the space in the main where the broken pipe was removed. Once it's lined up correctly, slide the two rubber couplings over the two standing ends and tighten the couplings to complete the repair. This method works on rigid coupling asbestos cement pipe as well as rubber coupling and bell and spigot pipes.

Placing sewer pipe is a fairly easy operation. If you have problems joining the pipe, or with pipe breakage, call the pipe manufacturer's representative. He can usually give you the assistance you need to solve the problem.

Tips for a Well-Planned Operation

An efficient sewer laying operation requires planning and coordination. Have the equipment you need available and calculate the required trench width carefully. The grade dug by the trencher must be accurate. Don't leave shallow areas that require hand work. Your shoring crew must be able to move quickly enough to stay ahead of the grading operation. Make sure the men grading have the material they need close at hand. They have to be fast enough to keep ahead of the pipe layers. The pipe should be laid out along the trench correctly so the pipe layers don't have to wait for pipe.

Figure 30-6 shows a well-planned trenching and grading operation. The trencher has excellent crumbing shoes to keep the dirt pulled into the bucket line. The man doing the crumbing has no trouble keeping up with the machine. The man behind the bucket line is throwing loose material back into the bucket line so the bottom will be firm for the crushed rock. One man is checking grade for the man grading.

Many areas are now using plastic pipe for sewer lines. The big advantage of plastic pipe is that long lengths are light enough to be handled by one person. This is very helpful if the trench is heavily

A well-planned trenching and grading job
Figure 30-6

shored and the pipe must be maneuvered down through the shoring.

You should partially backfill the trench once your pipe is laid. The specifications may call for a sand or aggregate backfill of 1 foot over the pipe before native material can be used. If no initial backfill is required, be careful that no soil chunks fall on the pipe. A heavy rock or chunk of hardpan could break the pipe. Shove material into the trench at a 45-degree angle to the trench. This way the soil will tend to roll over the pipe rather than fall directly on it. Don't cover lateral tees if the laterals have not been laid yet. Mark them with a stick or lath and shove a small amount of backfill material over the tee.

Check your specifications for compaction requirements. If water jetting is allowed, be sure the jet rod is not so long that it will hit and possibly break the pipe.

If the pipe being laid is too large to lay by hand, then you must use a crane. Truck mounted cranes are excellent for laying sewer line because pallets of pipe can be carried on the truck bed. The crane can lay each pipe directly from the supply on the bed.

Remember, it's important to check each pipe for slope carefully. An indication of a poorly laid sewer line is water standing in the pipe. You can see this at the manholes.

Laying Drain Pipe

As with water and sewer pipe, the first step in laying drain pipe is to prepare for delivery of the pipe to the job site. Be sure you have the correct equipment on hand to unload the pipe. You'll need to know the length and weight of the pipe that will be delivered, and whether the pipe will be banded on pallets or in single lengths. Plan to use slings or pipe hooks to help avoid damage to the pipe. Many pipe manufacturers will supply the equipment and do the unloading for you.

Setting Up the Worksite

Regardless of who unloads the pipe, it's extremely important to place the right amount of pipe where the laying crew needs it. To find the number of pipe that should be placed between grade stations, divide the length of pipe into the footage between stations. The answer is the number of pipe sections you need in that area. If the pipe isn't distributed correctly, your production will be greatly slowed.

Swing crane lowers pipe into trench
Figure 31-1

When unloading the pipe, determine the distance between the pipe and the trench by considering the type of equipment you'll use to lay the pipe. If you're using a swing crane, place the pipe far enough from the trench to allow driving room between the trench and the pipe. This gives the operator a better view of the men in the trench without the pipe blocking his vision. It also leaves more room to place trench shoring without working around pipe. And gravel can be dumped in the trench for grading without working over the pipe. Figure 31-1 shows the space available for working around the trench when you use a swing crane. But if you're using a rigid side crane that can only raise and lower loads, the pipe must be unloaded along the trench edge.

Once the pipe is unloaded, dump any trench material you need in strategic locations along the trench. Compute the amount of bedding material and initial backfill you'll need in advance so you can have it placed between stations or at every other station. If you're using cement type pipe joints, sand and cement must be dumped in several locations along the trench as well.

Grading the Trench Bottom

If you use a bottom grade string or laser, two people can handle the trench grading. If grade is established from a top line over the trench and if the string line or laser is higher than 8 feet, have a third person checking grade to speed up the operation. If aggregate for the trench bottom is being dumped in the trench with a hoe or loader, use a second crew member in the trench to spread the aggregate ahead of the one who's grading. Someone at the top of the trench can check the grade and direct the hoe or loader.

If there is excess water at the bottom of the trench, undercut the final grade enough to get down to firm soil, or excavate deep enough so that you can bridge the area with crushed rock. Crushed rock is a good bedding material. It stays firm even when water is running through or over it. Never attempt to lay pipe on a trench bottom that feels spongy underfoot.

In a well-balanced pipe laying operation, you need a crew that's just large enough to keep up with the trenching operation. There may be times when working around utilities slows the trenching operation to the point where a crew of three can handle the shoring, grading, laying, banding and backfilling of the pipe, and still keep up with the trencher. The footage you lay each day can vary greatly if existing utilities are a problem.

Types of Drain Pipe

There are three types of drain pipe in general use, and each takes a slightly different method of laying. The three types are shown in Figure 31-2.

In some areas, plastic drain pipe is being used. This pipe is light and easy to handle. It's thin-walled, with many ribs for added

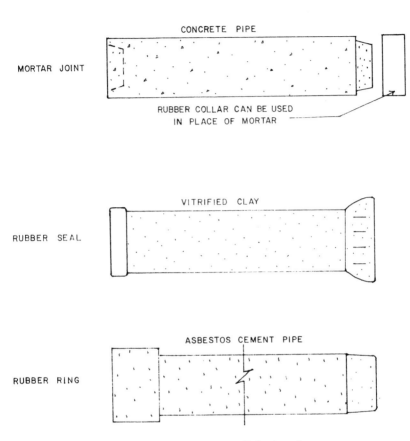

Three common types of drain pipe
Figure 31-2

strength. Figure 31-3 shows a 13-foot section of 18-inch plastic drain pipe being rolled to the trench by one man. One end is belled, and there's a rubber ring on the other end for easy installation.

Slip Ring Pipe
Slip ring drain pipe, like all drain pipe, should always be laid from the downstream end to the upstream end of the trench. Be sure you lay the pipe bells in the correct direction. They should face

Excavation & Grading Handbook

Plastic drain pipe is light weight
Figure 31-3

upstream. Join the rubber ring type drain line by lubricating the bell and spigot of the pipe and pushing the two together. A steel bar jabbed in the ground can be used as a lever to force them together. Be sure that no dirt or gravel gets into the collar.

You must line up each pipe with the pipe it's joining. If they aren't straight, joining the two will be difficult, if not impossible. Forcing two pipes together that aren't properly lined up may cause the rubber ring to roll from its groove and become crimped. This will result in a leak. Even if you're laying on a radius, you must still line the pipe up straight to join it. After it's seated, *then* turn it to make the radius.

Rubber Collar Pipe

You lay the rubber collar concrete pipe in the same manner as the ring type pipe except that no lubricant is used before joining the pipe. Place the rubber collar on the pipe in the trench, then shove the pipe being laid into the collar. Be sure to keep the joints free of dirt so they'll join completely.

Cement Joint Pipe

Laying pipe that requires a cement joint is much more difficult than laying rubber joining pipe. The pipes are joined together with mortar and then banded or sealed with additional mortar. You need one person just to keep a supply of mortar available for the laying and banding operation.

The size of the mortar mixing equipment depends on the size of the pipe and how many feet a day you'll be laying. For example, if the total production for the day will be about 200 feet and the pipe is 12 inches in diameter or less, you can use a wheelbarrow for mixing mortar. If 400 feet of pipe will be laid and the pipe is 12 inches in diameter or less, use a mortar box for mixing. If the pipe is 36 inches in diameter and 400 feet will be laid each day, then you'll need a gas-powered mortar mixer.

Laying Cement Joint Pipe

Laying concrete pipe in a high production operation takes several people. You need two on top, one to hook the pipe and one supplying mortar. In the trench you need two crew members laying pipe and one banding. When you lay 48 inch or larger concrete drain pipe, you may need two for banding, one on each side of the pipe. An additional two members work ahead of the pipe laying operation grading the trench. The first one grades and the second stays on top and supplies or guides grading material to the man in the trench. He may also check grade for him.

Assume that the grading has proceeded far enough down the trench for the pipe laying to begin. The first step in laying cement joint pipe is to mix the mortar. You need a creamy consistency so the mix will stick to the pipe. The water content is very important.

If the mortar is too soft, it will fall off the pipe. If it's too stiff, it will be hard to apply.

The following is a good rule of thumb when mixing mortar: Take a shovel full of mortar. Hold it 3 feet above the mortar box and let it slide off the shovel. If it disappears well into the surface of the mortar below, it's too soft. If it stays in a mound on top of the mortar, it's too dry. If it only partly submerges, it should be good. A good sand-to-cement ratio to use is one sack of cement to 20 square-nose shovels heaped with sand, and one shovel heaped with mortar cream. If you use plastic cement, you won't need the mortar cream.

Use a 5-gallon bucket, about two-thirds full, to pass the mortar to the pipe layer and his helper. The one or two people doing the banding will also need buckets of mortar and occasionally a half-full bucket of water passed to them. Anyone working with mortar, since they apply the cement with their hands, must wear rubber gloves to avoid cement poisoning.

The pipe layer should have a steel bar 5 or 6 feet long and about $1\frac{1}{8}$ inches in diameter. He also needs a shovel and a pole as long as the full length of the pipe, with a concrete brush nailed to it. The pipe layer's helper and the people banding the pipe also need concrete brushes, but with regular handles.

Lift the first length of pipe over the trench with a cable, preferably one connected to a quick coupler. The hookup man signals to the crane operator to tell him when to stop lowering the pipe. Set the first joint of pipe into place and then unhook it. The grooved or bell end should be facing upstream.

Dig a small depression for the bell in front of the pipe. The hole should be the width of the portion of pipe that's touching the ground. Put one or more handfuls of mortar in the small trench. The amount of mortar depends upon the size of the pipe. Using the short-handled concrete brush, splash water in the pipe groove. Then, smear mortar in the groove on the bottom half of the pipe. Now the buckets and tools must be moved back to make way for the second pipe.

Always check the trench grade where the next pipe will be set and make sure it's still level. If the grade has been disturbed from laying the first pipe, smooth the trench bottom with a shovel. When you're laying large pipe, it's especially important to keep a good grade slope on each pipe. Just a slight difference in grade slope between pipe will make a good joint impossible.

The second pipe should be hooked up and ready by this time. Lower it into the trench to about waist height. One man will need to splash water on the spigot end of the pipe and apply mortar to the top half of the spigot. Now lower the pipe until it's touching the trench bottom lightly. Keep just enough weight on the hoist line so that the pipe can be moved by hand.

Guide the spigot end of the second pipe into the groove end of the first pipe. Join the two by pushing the second pipe firmly into the groove of the first pipe using the steel bar for leverage. The two pipes must be lined up straight with each other or the groove and spigot won't slide together easily. If you're laying pipe on a curve, shove the pipe together first and then turn it to match the radius.

When the joints have gone together correctly, there will be a space between the two pipes. The gap will be approximately 3/8 inch between pipe up to 12 inches in diameter and 1 to 2 inches between pipe 48 to 60 inches in diameter. In a properly aligned joint, mortar will squeeze from between the two pipes as they are shoved together. If the cement is too dry, the pipes will resist being pushed together. When the pipes are firmly in place, check the flow of the pipe with a torpedo level, grade line, or laser. If the flow looks good, detach the cable sling from the pipe.

One man removes the excess mortar at the joint inside the pipe with the brush on the pole. He smooths off the mortar by using a sweeping motion around the inside of the joint. Any mortar left in the pipe should be brushed away. At the same time the other man smooths away the mortar that squeezed from the joint on the outside. This procedure is used for each new section of pipe that you lay.

Banding the Joints

The man or men banding the joints follow close behind the laying crew. They should dampen the pipe on the outside by splashing water on it before applying the mortar. If the pipe is small enough to straddle, they can take a handful of mortar in each hand and force it under the pipe. With each hand they'll apply a strip of mortar upward toward the top of the pipe. If the strips of mortar fall short of meeting each other at the top of the pipe, apply more mortar using the same procedure until the pipe has been banded all around. Jamming the mortar under the bottom side of the pipe joins it with the mortar the pipe layer placed in the small trench under the pipe while laying the joint. This makes a complete mortar band around the joint.

The thickness and width of the band will vary with the size of the pipe. For example, 12-inch diameter pipe should have a band approximately 4 inches wide and 1/2 inch thick. A 60-inch diameter pipe needs a band approximately 7 inches wide and $1\frac{1}{2}$ inches thick. After applying the band, smooth it with a damp brush. If the pipe laying is proceeding slowly for one reason or another, the banding can be done by the pipe layer's helper after each section is laid. The banding operation should stay a few joints behind the pipe laying so that the banded joints are not jarred as the pipe is pushed together.

Your job specifications may call for a cover of some type to protect the bands from drying too quickly. The most common covers are sprinkling dirt over the bands, brushing paper over the bands to protect them from the air and sun, or spraying a curing compound on them.

Backfilling the Trench

Once the pipe has been laid and the bands have dried sufficiently, you can backfill the trench. The initial backfill must be done cautiously. A sudden surge of material on one side can cause the pipe to roll and crack the bands. If your job specifications require that the backfill be tamped in layers, they should also specify that the bands have time to cure before any compacting is done. Curing is also necessary before water jetting. If no time is specified, you should wait at least 8 hours before jetting the fill.

If concrete pipe has to be cut for any reason, a carpenter's hammer may do the job. Use the hammer to chip around the pipe

Rubber ring type pipe
Figure 31-4

several times, creating a deeper groove each time, until the pipe is finally cut through. The size of the hammer depends on the thickness and size of the pipe. You may need a jackhammer and a chisel on very large pipe. When you're using a hammer to cut pipe, be sure to wear safety glasses.

Pipe can also be cut with a power saw with a masonry blade. And there are pipe cutters for cutting most cast iron, asbestos, and vitrified clay pipe.

When you're laying and banding pipe, plan your mixing so the last batch of mortar is exhausted before the end of the shift. Your crew will need time for cleaning the mixing equipment, buckets, and gloves before quitting.

Alternatives to Concrete Joint Pipe

Concrete joint pipe is a lot of work to band. That's why ring type and rubber collar type pipes are being used more often. Laying concrete ring type pipe is a much easier and cleaner job than mortaring joints. Figure 31-4 shows a rubber ring pipe being swung into the trench. Notice the ring laying inside the pipe. The pipe layers need only to stretch the ring into the spigot groove before joining the pipes. It takes just three people to lay this type of pipe.

Plastic drain pipe is getting more popular each year. It's light and easily joined with rubber rings. Corrugated steel and aluminum pipe with bolt-on bands are used frequently on road jobs. Concrete pipe that must be mortared at the joints is becoming less popular because of the extra labor cost for banding.

Constructing
Manholes

I explained how to excavate for manholes in an earlier chapter. In this chapter I'll explain how to build up the concrete manhole form and set the grade rings that level the manhole cover with the finished grade. Figure 32-1 shows the poured concrete bottom, precast concrete barrels and dome, and the location of the castings and grade rings.

Unless you have a sump at the bottom, you need to build manholes that allow for a smooth flow of water. This is more easily accomplished when you lay the pipe through the manhole bottom. If a side lateral enters the manhole, lay it into the manhole to a point where the sweep in the poured bottom starts to enter the main line. See section A in Figure 32-2. If a second lateral enters the manhole from the opposite side, lay it the same way. This will allow a good flow from the side channels into the main. You'll need to plug the side lateral pipes with sand bags or something similar to keep the concrete from running into them when you

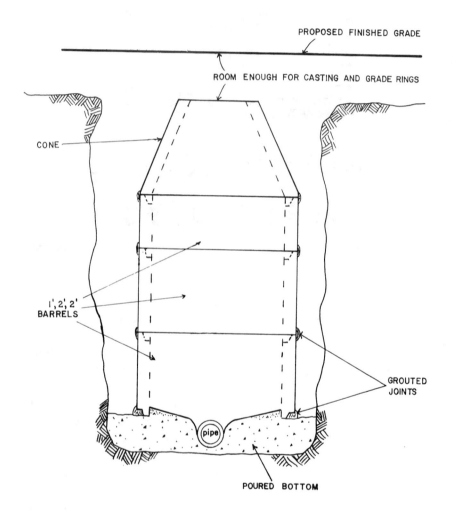

PROPOSED FINISHED GRADE

ROOM ENOUGH FOR CASTING AND GRADE RINGS

CONE

1', 2', 2'
BARRELS

GROUTED
JOINTS

pipe

POURED BOTTOM

Manhole
Figure 32-1

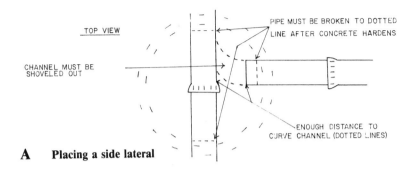

A   Placing a side lateral

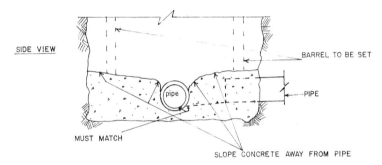

B   Initial shaping

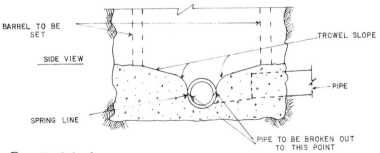

C   Final shaping

Poured manhole bottom
Figure 32-2

pour the bottom. Be sure that the manhole is excavated far enough under the pipe, and is the required diameter to comply with your job specifications.

Pouring a Manhole Bottom

When pouring the manhole bottom, be careful you don't pour concrete directly on the pipes that extend across or into the manhole. If you must set the manhole barrels soon after pouring the bottom and the weather is cool, use 2 percent calcium chloride in the concrete mix for a quick set. Take care to keep the pipe from floating as the concrete rises under it. A dry mix, around a 3-inch slump, will help eliminate this problem.

Pour the concrete to about 2 inches above the top of the pipe, and then begin to slop and channel the bottom. See Figure 32-2 B. After some initial shaping, let the concrete dry until it's firm enough to hold the shape you want. Then do the final shaping and troweling. It's important to shape the concrete wall at the spring line of the pipe so the bottom has a full channel width, as shown in C in Figure 32-2. The channel should always be as wide or wider than the pipe at every point. If you're also channeling in side laterals, be sure the concrete has a good sweep and is sloped low enough to allow an even flow into the main line.

Don't break out the pipe laid through the manhole until the concrete bottom has hardened for at least 12 hours. If the concrete isn't allowed enough curing time before the pipe is broken, the pipe may crack past the poured bottom back into the line. This will cause a leak, and you'll have to replace the pipe.

After you break away the excess pipe, grout the manhole bottom to give it the appearance of being molded into shape. Figure 32-3 is a precast molded bottom. A hand finished bottom will not look quite this smooth, but you should try to make it as much like this as possible. You need the following tools for finishing a manhole bottom: a hammer, a square nose shovel, rubber gloves, and a finisher's or brick layer's trowel. Once you've troweled the bottom as smooth as possible, use a fine brush to lightly etch the finished surface.

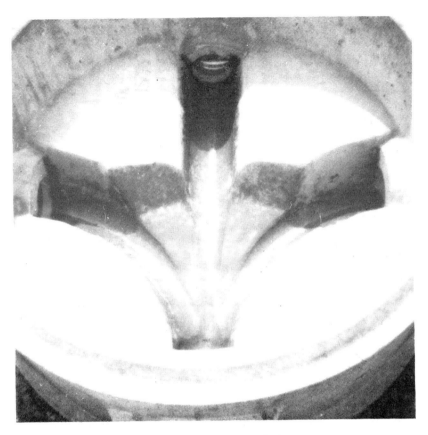

A molded precast manhole bottom
Figure 32-3

Precast Manhole Bottoms

If precast manhole bottoms are available, you should consider using them. They are much faster to set. See Figure 32-4. Overexcavate for the manhole bottom. Then, allowing for the thickness of the precast bottom, place a level layer of gravel in the excavation. Make sure that when the bottom is set on the gravel bed, the pipe in the trench will match the openings for the pipe of the precast bottom. Whether the bottom is poured or precast, it must be level so the manhole won't lean when the barrels are set.

Precast manhole bottoms save time
Figure 32-4

When the pipe is laid up to the manhole, lower the precast manhole bottom into place. Connect it to the downstream end of the sewer with a short section of pipe that has two male ends. Once in place, lay the upstream pipe into the manhole. When the pipe is joined to the manhole, the main line work can continue. The manhole bottom is now finished and ready for setting the barrels.

A precast manhole bottom is ideal when time is important because it eliminates the wait of two or more hours between pouring a concrete bottom and setting the barrels.

Pouring a Manhole with a Sump
When specifications require a sump, you shouldn't lay the pipe through the manhole. Instead, extend the pipe just far enough into the manhole to reach the inner wall. After digging the hole to the depth specified, pour the bottom. Next, insert the inner wall forms and set them on the concrete bottom.

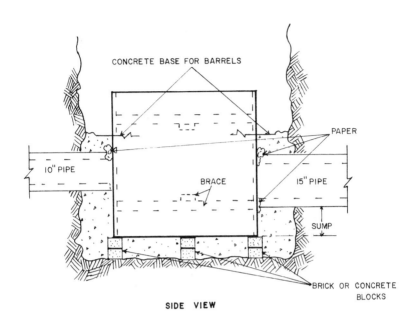

CONCRETE BASE FOR BARRELS

PAPER

10" PIPE

BRACE

15" PIPE

SUMP

BRICK OR CONCRETE BLOCKS

SIDE VIEW

Form for a manhole with sump
Figure 32-5

You can also raise the inner wall forms to the bottom grade as shown in Figure 32-5. Set them up on brick or concrete blocks. Then you can pour the walls and bottom at the same time. Use wood forms or Sonotube for the inner wall forms. Trim the pipe to length carefully with a hammer or pipe cutter so the form has just enough room to slide between the two pipe ends. See Figure 32-5. If there is a space between the form and pipe large enough for the liquid concrete to pass through, use newspaper or paper from concrete sacks to plug the holes. Place braces inside the form to keep it from distorting from the weight of the concrete.

The sides of the manhole should be poured first if both the bottom and the sides are being poured. Pour several inches of concrete all the way around the form. Never let the concrete build up on one side of the form. The weight of the concrete may cause the form to slide off center. Repeat the circular pouring pattern until the concrete reaches the top of the pipe. Always pour to the top of the highest pipe if the pipes entering the manhole are at different elevations. Check across in both directions with a level on a straight edge to make sure the manhole remains level.

Use a pole to prod and settle the concrete, making sure that all the voids are filled. Tap the inside of the form slightly with a hammer as the concrete is being poured. This helps make the surface smoother after the forms have been pulled.

When you pour small manhole bottoms without outside forms, don't use a concrete vibrator. A vibrator would cause the concrete to seep into the pipe at the end spaces.

Once the sides have been poured, pour the bottom. Be careful to keep the concrete just below the bottom form. If the concrete is poured higher than the bottom edge of the form, the form will be very hard to remove once the concrete hardens.

The time needed for curing before the forms are pulled depends on the weather. On a warm day, with calcium chloride added to the concrete, three hours curing should be enough if time is important. To test the concrete, bang it with the handle of a shovel. If the handle sinks in or dents the concrete easily, don't pull the forms. If the concrete feels solid, the forms can be removed. Once the forms have been pulled, clean the paper from the ends of the pipe and knock off any rough points of concrete. The manhole can be grouted with a mixture of sand and cement to fill any voids. Finally, brush the concrete smooth.

Setting the Barrels

Now you're ready to set the precast barrels. Your job specifications will state the type of joint material you need. Usually cement or a tar compound is required. Setting precast barrels is like stacking blocks; most have a tongue and groove. Place the tongue down and the groove up. The barrels come in various lengths. Choose the correct length for each manhole so that the top barrel will correspond with the finished ground level as shown in Figure 32-1.

If the street or parking area has not been cut to subgrade elevation, have the surveyors set the finished grade elevation for the top of the manhole. If the surveyors aren't available, get the elevation from the plans.

The plans should give the invert elevation (or lowest level) of the pipe at the manhole, and also the street or ground elevation. Subtracting the invert elevation from the street elevation will give you the total height of the manhole from the flow line. Now subtract from that figure the thickness of the cast iron manhole casting for the top, and the distance from the flow line to the top of the poured bottom. This gives you the required height of barrels and cone.

You can make any minor adjustments with 3-inch or 6-inch grade rings. Your job specifications will list a maximum number of grade rings that you can use on each manhole. Be sure the barrels are set high enough to keep the number of grade rings within specifications.

Setting the Manhole Casting

The generally accepted way of placing manholes in new streets is to leave them deep enough so that they can be paved over. After the paving has been completed, the manholes are uncovered and grade rings and castings are set.

You must be sure, when setting the castings, that they are level with the pavement. With wire, tie 2 x 4 boards to the manhole casting and hang it from the pavement over the manhole. Then set the grade rings and pour the concrete around the casting. See Figure 32-6. After the concrete has set up so it feels hard, cut the wire and remove the 2 x 4's. The wire left sticking slightly out of the concrete should be bent over with a hammer before paving around the manhole. After the manhole is paved, be sure that all the inside rings are grouted and the bottom cleaned.

If you're working on a traveled street where the raised casting must be poured and paved around, timing is important. Still, you must be careful when paving around manholes. If you don't make a smooth match to the existing pavement, it will be obvious to every driver on the roadway. Use 1/2 or 3/8 inch aggregate mix.

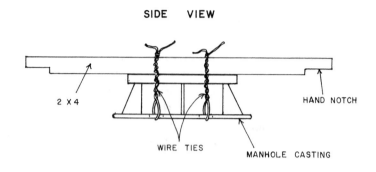

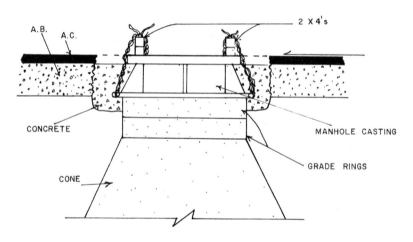

Setting manhole casting
Figure 32-6

Be sure to tack coat the bottom and edges before paving. A good four-man crew can finish digging, setting, and paving nine manholes in an eight-hour day.

A word of caution: While setting manholes and handling manhole lids, be sure fingers are kept clear. And always wear rubber gloves when grouting or working with concrete to avoid concrete poisoning.

33

Pressure Testing Sewer Pipe

There are two ways to test sewer lines for leaks: water tests and air tests. Most specifications require that sewer lines be tested after installation and backfilling are completed. If house laterals are already connected to the new house service lines, they must be plugged before this testing is begun.

The Water Test

To make a water test for gravity sewer lines, the section being tested must be plugged at each end. Plug both the downstream and upstream manholes on the upstream side. See Figure 33-1. Any pipe entering the upstream manhole from the sides must be plugged unless it's a service line. Service lines must be included in the test. It's good practice to tie a wire or short length of rope to each plug and wrap the wire or rope around a small board. This

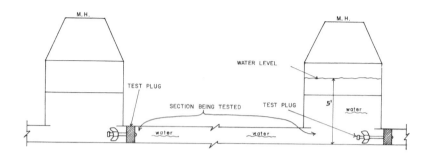

Water test for sewer pipe
Figure 33-1

will keep the plug from being pushed down the pipe by the water pressure if the plug comes loose.

Once the plugs are secure, release water into the upstream manhole until the pipe is full and the water rises about 5 feet up the manhole barrels. Most specifications require a minimum height of 5 feet above the pipe. This supplies enough pressure to the main to conduct the test. Four hours should elapse before testing begins so that the pipe and manhole are fully saturated. After four hours, add water to replace the water that was absorbed into the pipe and manhole walls. Once the water is again 5 feet above the pipe, you may begin the test. The time usually required for a sewer test is 2 to 4 hours.

The amount of leakage you'll be allowed may vary from one job specification to the next, depending on the agency involved. The leakage tolerated is usually 250 to 500 gallons per mile per 24 hours per inch of diameter of the pipe tested. The leakage tolerated in a sewer force main is much less, usually around 125 gallons per 24 hours.

Don't fill the manhole with more water than required by the specifications. Each extra foot of water increases the pressure in the main substantially.

It is to your advantage to test before all the backfill material has been placed, especially if the trench is very deep. You should build manholes as soon as possible so you can begin testing.

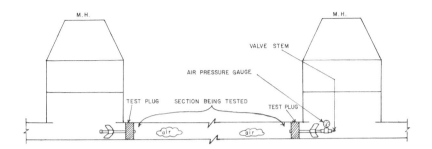

Air test for sewer pipe
Figure 33-2

If a leak develops while testing, keep a good head of water in the manhole until the water seeps to the top of the ground. If you have only placed a small amount of backfill, this will occur much sooner and will help you pinpoint the leak faster. If the specifications require that you completely backfill and jet or compact the sewer line before testing, you would be wise to conduct your own test before backfilling and compacting. It's very hard to find a leak during a water test when the trench is deep and it's been jetted or compacted. Remember, if you have a leak, always check the manhole joints for cracks. It may be the manhole that's leaking, not the pipe.

The Air Test

The air test is the fastest, cleanest, and least expensive way of testing sewer mains and services. The upstream manhole must be plugged on the downstream side in this test. The downstream manhole is plugged as in a water test. This way the air is trapped between the manholes' openings. One of the two plugs set must have a pressure gauge and a valve stem as shown in Figure 33-2. It's not necessary to tie the plugs to a board when air testing.

Once the plugs are set, air is pumped into the line. An air nozzle

Excavation & Grading Handbook

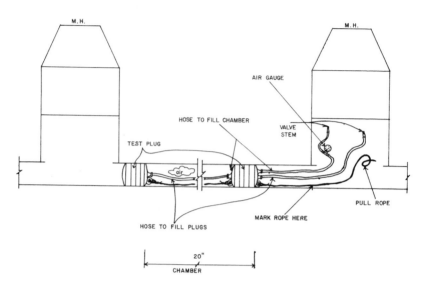

Locating leaks
Figure 33-3

and compressor are needed to fill the line with air. Most agencies require that 3 to 5 pounds of pressure be pumped into the main and services. Usually the pressure must be held for 5 minutes with a maximum pressure loss of 0.5 pound per square inch. The time the pressure must be held varies with the size of the pipe being tested. Your job specifications will list the amount of time and pressure required.

The big advantage of air testing is that the line can be tested before the manholes are built. If you're laying pipe under adverse conditions and leaks are a concern, test the pipe regularly during the laying operation.

Locating Leaks with the Air Test
When a leak is detected during the test, it must be located. Even if the main has been jetted or compacted, a leak is easily located if you have the correct apparatus. To locate the leak, remove the two test plugs. A special set of test plugs is used to pinpoint the leak.

These special plugs are inflatable and are connected by an air hose and a rope as shown in Figure 33-3. Measure the distance from manhole to manhole and attach enough air hose and rope so that the test plugs can pass through the entire section to the downstream manhole.

To get the rope and air lines through the pipe, a small cord is floated down through the line. The end of this cord is connected to the plugs, and the rope and air lines are pulled back through the line to the downstream manhole. The cord must be pulled until both plugs on the test apparatus enter the pipe at the upstream manhole. One man must be in each manhole until the plugs and line are in place in the main. Once the second plug enters the pipe, locating the leak may begin.

Make a mark on the rope where it enters the pipe at the downstream manhole. Marking the rope is necessary to determine how far the test apparatus has been pulled down the line once a leak is found. The air hose fills both plugs with air to seal off each test section. Usually 20 feet are tested at a time. Once the plugs have been inflated, air is pumped into the second line to pressurize the pipe section between the two plugs. Fill the pipe section between the two plugs to the air pressure specified and wait two minutes. If the pressure holds, release the plug pressure. This also releases the pressure in the pipe. Once the pressure has been released, pull the test plugs ahead 19 feet. Repeat this test cycle until you locate the section that won't hold pressure.

Now you have isolated the leak within a 20 foot section. To further pinpoint the leak, move the test plugs ahead 3 to 5 feet each time and retest. Continue this operation until the test pressure holds well again. Now measure the amount of rope that has been pulled from the pipe. If 100 feet has been pulled through, measure 100 feet from the downstream manhole toward the upstream manhole and mark the ground at that point. This will give you the location of the last plug on the test apparatus and will pinpoint the leak to within 3 to 5 feet, the distance the plugs were moved ahead each time.

Continue moving the plugs 19 feet each time until the entire main has been tested for leaks. There may be more than one leak, so don't discontinue testing when the first leak has been located.

After the leak has been pinpointed to within 3 to 5 feet, check the location of the sewer services. If a service enters the main at that location, the leak could be in the service line rather than the main line.

The best procedure to follow in this case would be to expose and check each end of the service line. Check the service where it connects to the main line and at the property line. These are the most likely spots for leaks. If there isn't a leak at either end, the entire length of the service line must be checked. Place test plugs at each end of the main and keep pressure in the line. The line should remain under pressure while you excavate to make finding the leak easier.

The leak may be caused by a stone or dirt wedged in the rubber fitting. In that case, air rushing from the joint will pinpoint the problem area for you. If the leak is caused by foreign matter in a joint, you may be able to clear off enough room to pull the pipe apart, clean it, and buckle it back together.

If the problem is a cracked pipe, buckle a new section of pipe into place. If the line is too large to buckle a new section in, or if there isn't enough room to do it, you must cut out the cracked section and drop in a new one. Use rubber caulder coupling to join each cut end.

Correcting leaks is expensive. A little care while laying the pipe will reduce the number of leaks and save a lot of time and money.

34

Drains and Culverts

Several types of drains are used to keep water from accumulating on or under roadways. This chapter covers underdrains, culverts, and downdrains.

Underdrains

Underdrains are designed to collect and carry off water that accumulates under the road surface. This is done by laying a pipe with several holes drilled at the bottom in a shallow trench filled with a permeable material. The permeable material is usually rock with only a small amount of sand or fines, so the water can seep freely through the rock.

Fill the ditch with rock from 6 inches under the pipe to the top of the ditch. Then make sure there's no dirt or subgrade material between the rock base used on the road and the ditch rock. To accomplish this, you'll have to trim and roll the subgrade of the road and get it accepted by the inspector, before you can install the underdrain.

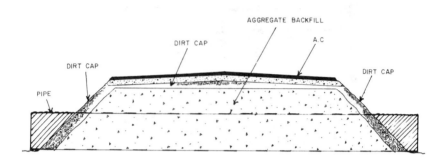

Culvert with dirt cap
Figure 34-1

When the inspection is completed, dig the underdrain pipe trench. All dirt from the trench must be hauled off. Spread 6 inches of rock under the pipe, lay pipe in the trench, and then fill the trench to subgrade level with more rock. Spread the road aggregate when you complete backfilling the trench.

You usually place underdrain lines at the outside edge of the pavement or shoulder. They can be at the high side or low side of the road, depending on where seepage is expected.

Culverts

One good way to control surface water is with a culvert. The engineer may design a culvert with a dirt cap (Figure 34-1) so the road rock and the underdrain rock don't come together, as they do in an underdrain. If you need a dirt cap, lay the culvert pipe after you complete the rough subgrading of the road. Culvert drain pipe isn't perforated, so the water flows away through the pipe. Concrete or corrugated drain pipe is usually used for culverts.

Downdrains

Downdrains are designed to handle water runoff from the road and shoulder surface. The three most common downdrains are

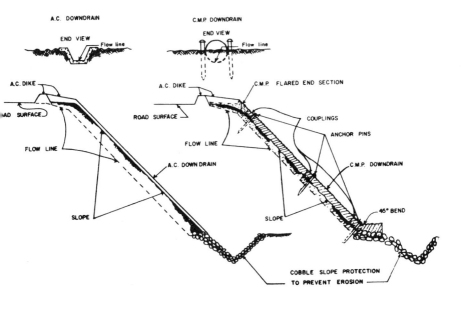

Downdrain detail
Figure 34-2

corrugated metal pipe, corrugated metal trough, and asphalt pavement. Here's the main point to remember about placing downdrains: In most cases they can't be placed until the road is paved and the shoulders have been finished or paved. Putting them in earlier will obstruct and slow the work.

You'll usually need to put in a dike with a downdrain. The drain is put in first, then the dike. If the downdrain is an asphalt trough, you can put it in during the dike operation. If you use metal pipe for the downdrain, anchor it with metal stakes. Downdrain pipe is usually laid very shallow with the top of the pipe exposed. If the slope is long, brace the pipe to keep it from sliding. Figure 34-2 shows both an asphalt trough and a metal pipe downdrain.

Some downdrain outlets are designed to run vertically deep into the fill slope and out of the bank well below the road surface. In this case, the drain will have to be put in as the fill slope is built up. Cap the pipe just below the shoulder grade and then uncover and finish it after the shoulders are trimmed. In some cases you can extend the drain pipe up a section at a time during the fill operation.

A metal trough downdrain is primarily a surface-type drain. Be sure to anchor the drain to keep it from sliding down the slope when it fills with water. You can install this type of drain just ahead of the dike operation.

Placing Drains and Culverts

Your main concern in placing drains and culverts is timing. If you put them in too late, you may have to dig through 10 feet of fill to do it. If they're placed too soon, they may be damaged by the excavating equipment. It takes judgment and experience to know when the drain or culvert should be placed. There's no single time during the excavation work that suits all situations.

The best time may depend on the depth of the culvert. For example, if you need to build the fill up 60 feet high and the plans show a culvert 5 feet in diameter running under the fill, the most common method is to build the fill above the culvert to be placed. For a 5-foot diameter culvert, build the fill to a point approximately 2 to 3 feet above the top of the culvert. Then lay the culvert before building the remainder of the 60-foot fill. You may elect to build the fill just 4½ feet high so no trench shoring is needed. This is 6 inches shy of covering the culvert. So when you continue the fill, you'll need to build a ramp over the culvert. The dirt ramp must be thick enough to prevent damage when equipment passes over the pipe.

On many jobs the drain lines and culverts are just below subgrade. Do the rough grading and the finished subgrade work *before* you lay the drains. If the drains are shallow, haul away the material excavated from the ditch, or use it for road fill. You need to use gravel to fill the ditch and backfill the pipe. Keep some dirt fill to plug the ends and cap the top of the drain, as shown in Figure 34-1. You may need to encase a shallow pipe in concrete if it's required in the specifications.

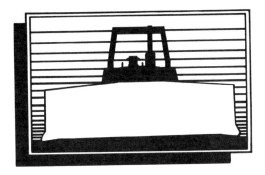

Glossary
and
Abbreviations

Backhoe: Self-powered excavation equipment that digs by pulling a boom mounted bucket toward itself. It also has a front bucket. (See Figure 27-8.)

Balancing subgrade: Trimming subgrade until there are several areas which are still too high or low, but which, when fine trimmed, will average out to the finished subgrade tolerance required.

Bank plug: Piece of lumber (usually 2" x 4") driven into the ground to stand some distance, usually 24", above ground level. Surveyors place nails in the bank plugs a given distance above the road surface so a string line can be stretched between the plugs to measure grade. (See Figure 10-8.)

Bench mark: Point of known elevation from which the surveyors can establish all their grades.

Benching: Making steplike cuts into a slope. Used for erosion control or to tie a new fill into an existing slope.

Bitch pot: Name used for an oil pot when it contains asphaltic emulsion (oil mixed with water), for a tack coat. It's usually trailer-mounted and pulled by a truck.

Bones: Rocks in the aggregate base which have come to the surface and separated from the finer material. Such a surface is called a "bony" grade.

Boot: A lath set behind the hub by the grade setter when there are obstructions blocking the line of sight to the hub. The grade setter draws a horizontal line on the lath 1 foot or more above the hub and shoots grade from this line.

Boot truck: Another name for an oil truck with a spray rack for spraying asphalt oil.

Borrow site: An area from which earth is taken for hauling to a jobsite which is short of earth needed to build an embankment.

Catch basin: A complete drain box made in various depths and sizes. Water drains into a pit, then from it through a pipe connected to the box.

Catch point: Another name for hinge point or top of shoulder.

Center line: The point on stakes or drawings which indicate the half-way point between two sides.

Chip seal: Process in which fine crushed rock is spread on asphalt oil and then rolled.

Chokers: Road shoulders that are to remain higher than the subgrade level. (See Figure 10-8.)

Clear and grub: To remove all vegetation, trees, concrete, or anything that will interfere with construction inside the limits of the project.

Compactor: A machine for compacting soil. It can be pulled or self powered. The latter have wheels to help compaction. (See Figure 7-4.)

Crows foot: A lath set by the grade setter with markings to indicate the final grade at a certain point.

Crumbing shoe: Metal arm-like attachment on wheel trenchers which keeps loose earth at the trench bottom pulled back into the digging buckets. (See Figure 27-1.)

Curb shoe: A device bolted to the blade of a grader when grading curbs. It helps the blade match the shape of the curb bottom.

Dikes: Raised sections built onto the sides of roads to control water runoff and erosion.

Disc: One or more rows of plate-shaped steel wheels, about 3/16'' thick, which cut into the earth, turning and mixing the soil.

Elevation numbers: The vertical distance above or below sea level.

Embankment: Area being filled with earth.

Feathering: Raking new asphalt to join smoothly with the existing asphalt.

Finished grade: Any surface which has been cut or built to the elevation indicated for that point.

Grade: The surface of a road, channel, or natural ground area. Usually means the surface level required by the plans or specifications.

Grade break: A change in slope from one incline ratio to another.

Grade lath: A piece of lath that the surveyor or grade setter has marked to indicate the correct grade to the operators.

Grade pins: Steel rods driven into the ground at each surveyor's hub. A string is stretched between them at the grade indicated on the survey stakes, or a constant distance above the grade.

Grade rod: A small length of round or rectangular wood or metal used in place of a ruler for checking grades.

Grader: A power excavating machine with a central blade that can be angled to cast soil on either side. It has an independent hoist control on each side. Also called a "blade." (See Figure 10-5.)

Guinea: A survey marker driven to grade. It may be colored with paint or crayon. Used for finishing and fine trimming. Also called a *hub.*

Guinea hopper: A member of the grading crew who uncovers the hub and signals the blade operator to cut or fill as required. (See Figure 21-2.)

High centered: Condition in which the tracks or wheels of equipment sink into soft soil. The undercarriage resting on soil prevents the equipment from moving out of the soft area.

Hinge point: A point indicating where the fill slope stops and the road or shoulder grade begins. Sometimes called the "catch point."

Hoe: A track-mounted, self-powered shoveling machine that digs by pulling a boom mounted bucket toward itself. (See Figure 25-1.)

Hubs: Point of origin stakes which identify a point on the ground. The top of the hub establishes the point from which soil elevations and distances are computed, or a point to be trimmed to.

Hypochlorite tablets: Tablets placed inside each joint of water pipe for chlorination and purification of water.

Information stake: Explains in surveyor's code what grades are to be established and the distances to them.

Kicker blocks: Cement poured behind each bend or angle of water pipe for support. Also called *thrust blocks.*

Lane delineator: A cylinder approximately 3 feet high with a rubber base. Used in a series to control traffic.

Lift: Any layer of material or soil placed upon another.

Lug down: A slowdown in engine speed (RPM) due to increasing the load beyond capacity. Usually occurs when heavy machinery is crossing soft or unstable soil or is pushing or pulling beyond its capability.

Mat: Asphalt as it comes out of a spreader box or paving machine in a smooth, flat form.

Maximum density and optimum moisture: The highest point on the moisture density curve. Considered the best compaction of the soil.

Median: The unpaved section between two or more lanes down the center of a highway.

MEE pipe: Pipe that has been milled on each end and left rough in the center. MEE stands for "milled each end."

MOA pipe: Pipe that has been milled end to end. MOA stands for "milled over all" and allows easier joining of the pipe if the length must be cut to fit.

Moisture density curve: A graph plotted from tests to determine at what point of added moisture the maximum density will occur. (See Figure 18-1.)

Natural ground: The original ground elevation before any excavation has been done.

Nuclear test: A test to determine soil compaction by sending nuclear impulses into the compacted soil and measuring the returned impulses reflected from the compacted particles.

Oil pot: Small tank on wheels that can be towed. Has a compressor to supply pressure so road oil can be sprayed from it with a hose and spray nozzle.

Paddle wheel scraper: An excavating machine which uses a conveying device to dislodge soil and move it into the bowl. (See Figure 9-2.)

Pneumatic tired roller: A roller with rubber tires commonly used for compacting trimmed subgrade asphalt or aggregate base. (See Figure 18-4.)

Popcorn: A name given to open graded asphaltic concrete having 3/4" aggregate with very little fine material.

Pug mill: A rectangular box on wheels containing rows of power driven steel arms that churn dirt and a mixture (usually lime) as it is pulled along the ground.

Pumping: A rolling motion in unstable ground. Usually occurs when heavy equipment passes over.

Quarter crown: The area between the centerline and the curb or shoulder running parallel to it.

Raveling: A cumulative process in which the rock separates from the finer material on the road surface because of car and truck traffic or excess blading.

Right-of-way line: A line on the side of a road marking the limit of the construction area and usually, the beginning of private property.

Ripper: Teeth-shaped attachments added to equipment to dig through hardpan or rocky soil. (See Figures 24-1 and 27-3.)

RS: Reference stake, from which measurements and grades are established.

Sand cone test: A test for determining the compaction level of soil, by removing an unknown quantity of soil and replacing it with a known quantity of sand.

Scraper: A digging, hauling, and grading machine having a cutting edge, a carrying bowl, a movable front wall, and a dumping mechanism. (See Figure 5-7.)

Sheepsfoot roller: A compacting roller with feet expanded at their outer tips. Used in compacting soil (See Figure 11-1.)

Spoil site: Area used to dispose of unsuitable or excess excavation material.

String line: A nylon line usually strung tightly between supports to indicate both direction and elevation. Used in checking grades or deviations in slopes or rises.

Structure section: Includes all the road material placed from the subgrade level to the finished road surface.

Subgrade: The uppermost level of material placed in embankment or left at cuts in the normal grading of a road bed. This becomes the foundation for aggregate and asphalt pavement.

Summit: The highest point of any area or grade.

Super: A continuous slope in one direction on a road.

Swale: A shallow dip made to allow the passage of water.

Swedes: A method of setting grades at a center point by sighting across the tops of three lath. Two lath are placed at a known correct elevation and the third is adjusted until it is at the correct elevation. (See Figure 13-1.)

Tangent: A straight line from one point to another, which passes over the edge of a curve.

T-bars: T-shaped wood frames used in place of steel pins to support a string line over trenches.

Tied out: The process of determining the fixed location of existing objects (manholes, meter boxes, etc.) in a street so that they may be uncovered and raised after paving.

Toe of slope: The bottom of an incline.

Track loader: A loader on tracks used for filling and loading materials. (See Figure 25-2.)

"Typical" drawing: End section view of a street or highway usually showing half of the road if both sides are the same.

Vertical curve: Indicates a curvature in a horizontal line to a higher or lower elevation.

Vibratory roller: A self-powered or towed compacting device which mechanically vibrates while it rolls. (See Figure 11-4.)

Windrow: The spill-off from the ends of a dozer or grader blade which forms a ridge of loose material. (See Figure 21-3.) A windrow may be deliberately placed for spreading by another machine.

Abbreviations

AB	Aggregate base	HP	Hinge point
AC	Asphalt concrete	IC	Interconnect
ACP	Asbestos cement pipe	ID	Inside diameter
ASB	Aggregate subbase	INV	Invert
BC	Begin curve or back of curb	LBS	Pounds
BM	Bench mark	LF	Lineal foot
BSP	Black steel pipe	LS	Lump sum
BV	Butterfly valve	LT	Left
CB	Catch basin	LTB	Lime treated base
CF	Cubic feet	MH	Manhole
CIP	Cast iron pipe	OC	On center
CISP	Cast iron soil pipe	OD	Outside diameter
℄	Centerline	PB	Pull box
CMP	Corrugated metal pipe	PCC	Portland cement concrete
CP	Concrete pipe	PG	Projected grade
CTB	Cement treated base	PI	Point indicated
CU	Conduit	PL	Property line
CV	Check valve	PMP	Perforated metal pipe
CY	Cubic yard	PSI	Pounds per square inch
DBL	Double	PVC	Polyvinyl chloride plastic pipe
DI	Drop inlet		
DR	Driveway	R =	Radius
EA	Each	RCB	Reinforced concrete box
EC	End of curve	RCP	Reinforced concrete pipe
EG	Existing grade	RP	Reference point
EL	Elevation	RT	Right
EMB	Embankment	R/W	Right-of-way line
EP	Edge of pavement	S =	Slope
EXC	Excavation	SD	Storm drain
FD	Floor drain	SF	Square foot
FG	Finished grade	SG	Subgrade
FL	Flow line	SS	Slope stake or sewer service
FH	Fire hydrant	ST	Station
FS	Finished surface	SY	Square yards
GAL	Gallon	TBC	Top back of curb
GB	Grade break	TC	Top of curb
GD	Gutter drain	VB	Valve box
GP	Grade plain	VC	Vertical curve
GSP	Galvanized steel pipe	VCP	Vitrified clay pipe
GV	Gate valve	WSP	Welded steel pipe

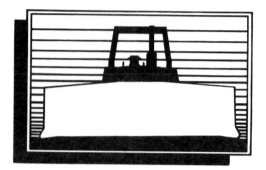

Index

Practical References for Builders

Basic Engineering for Builders

If you've ever been stumped by an engineering problem on the job, yet wanted to avoid the expense of hiring a qualified engineer, you should have this book. Here you'll find engineering principles explained in non-technical language and practical methods for applying them on the job. With the help of this book you'll be able to understand engineering functions in the plans and how to meet the requirements, how to get permits issued without the help of an engineer, and anticipate requirements for concrete, steel, wood and masonry. See why you sometimes have to hire an engineer and what you can undertake yourself: surveying, concrete, lumber loads and stresses, steel, masonry, plumbing, and HVAC systems. This book is designed to help the builder save money by understanding engineering principles that you can incorporate into the jobs you bid. **400 pages, 8½ x 11, $36.50**

Greenbook Standard Specifications For Public Works Construction

Since 1967, ten previous editions of the popular "Greenbook" have been used as the official specification, bidding and contract document for many cities, counties and public agencies throughout the West. New federal regulations mandate that all public construction use metric documentation. This complete reference, which meets this new requirement, provides uniform standards of quality and sound construction practice easily understood and used by engineers, public works officials, and contractors across the U.S. Includes hundreds of charts and tables. **730 pages, 5 x 8, $59.95**

Blueprint Reading for the Building Trades

How to read and understand construction documents, blueprints, and schedules. Includes layouts of structural, mechanical, HVAC and electrical drawings. Shows how to interpret sectional views, follow diagrams and schematics, and covers common problems with construction specifications. **192 pages, 5½ x 8½, $14.75**

Masonry & Concrete Construction Revised

This is the revised edition of the popular manual, with updated information on everything from on-site preplanning and layout through the construction of footings, foundations, walls, fireplaces and chimneys. There's an added appendix on safety regulations, with all the applicable OSHA sections pulled together into one handy condensed reference. There's new information on concrete, masonry and seismic reinforcement. Plus improved estimating techniques to help you win more construction bids. The emphasis is on integrating new techniques and improved materials with the tried-and-true methods. Includes information on cement and mortar types, mixes, coloring agents and additives, and suggestions on when, where and how to use them; calculating footing and foundation loads, with tables and formulas to use as references; forming materials and forming systems; pouring and reinforcing concrete slabs and flatwork; block and brick wall construction, including seismic requirements; crack control, masonry veneer construction, brick floors and pavements, including design considerations and materials; and cleaning, painting and repairing all types of masonry. **304 pages, 8½ x 11, $28.50**

Builder's Guide to Accounting Revised

Step-by-step, easy-to-follow guidelines for setting up and maintaining records for your building business. This practical, newly-revised guide to all accounting methods shows how to meet state and federal accounting requirements, explains the new depreciation rules, and describes how the Tax Reform Act can affect the way you keep records. Full of charts, diagrams, simple directions and examples, to help you keep track of where your money is going. Recommended reading for many state contractor's exams. **320 pages, 8½ x 11, $30.50**

Construction Forms & Contracts

125 forms you can copy and use — or load into your computer (from the FREE disk enclosed). Then you can customize the forms to fit your company, fill them out, and print. Loads into *Word* for *Windows*™, *Lotus 1-2-3*, *WordPerfect*, *Works*, or *Excel* programs. You'll find forms covering accounting, estimating, fieldwork, contracts, and general office. Each form comes with complete instructions on when to use it and how to fill it out. These forms were designed, tested and used by contractors, and will help keep your business organized, profitable and out of legal, accounting and collection troubles. Includes a CD-ROM for *Windows*™ and Mac. **400 pages, 8½ x 11, $41.75**

Contractor's Survival Manual

How to survive hard times and succeed during the up cycles. Shows what to do when the bills can't be paid, finding money and buying time, transferring debt, and all the alternatives to bankruptcy. Explains how to build profits, avoid problems in zoning and permits, taxes, time-keeping, and payroll. Unconventional advice on how to invest in inflation, get high appraisals, trade and postpone income, and stay hip-deep in profitable work. **160 pages, 8½ x 11, $22.25**

CD Estimator

If your computer has *Windows*™ and a CD-ROM drive, *CD Estimator* puts at your fingertips 85,000 construction costs for new construction, remodeling, renovation & insurance repair, electrical, plumbing, HVAC and painting. You'll also have the *National Estimator* program — a stand-alone estimating program for *Windows*™ that *Remodeling* magazine called a "computer wiz." Quarterly cost updates are available at no charge on the Internet. To help you create professional-looking estimates, the disk includes over 40 construction estimating and bidding forms in a format that's perfect for nearly any word processing or spreadsheet program for *Windows*™. And to top it off, a 70-minute interactive video teaches you how to use this CD-ROM to estimate construction costs. **CD Estimator is $68.50**

CD Estimator — Heavy

CD Estimator — Heavy has a complete 780-page heavy construction cost estimating volume for each of the 50 states. Select the cost database for the state where the work will be done. Includes thousands of cost estimates you won't find anywhere else, and in-depth coverage of demolition, hazardous materials remediation, tunneling, site utilities, precast concrete, structural framing, heavy timber construction, membrane waterproofing, industrial windows and doors, specialty finishes, built-in commercial and industrial equipment, and HVAC and electrical systems for commercial and industrial buildings. **CD Estimator — Heavy is $69.00**

National Construction Estimator

Current building costs for residential, commercial, and industrial construction. Estimated prices for every common building material. Provides manhours, recommended crew, and gives the labor cost for installation. Includes a CD-ROM with an electronic version of the book with *National Estimator*, a stand-alone *Windows*™ estimating program, plus an interactive multimedia video that shows how to use the disk to compile construction cost estimates. **560 pages, 8½ x 11, $47.50. Revised annually**

Pipe & Excavation Contracting

Shows how to read plans and compute quantities for both trench and surface excavation, figure crew and equipment productivity rates, estimate unit costs, bid the work, and get the bonds you need. Explains what equipment will deliver maximum productivity for a job, how to lay all types of water and sewer pipe, and how to switch your business to excavation work when you don't have pipe contracts. Covers asphalt and rock removal, working on steep slopes or in high groundwater, and how to avoid the pitfalls that can wipe out your profits on any job. **400 pages, 5½ x 8½, $29.00**

Estimating Excavation

How to calculate the amount of dirt you'll have to move and the cost of owning and operating the machines you'll do it with. Detailed, step-by-step instructions on how to assign bid prices to each part of the job, including labor and equipment costs. Also, the best ways to set up an organized and logical estimating system, take off from contour maps, estimate quantities in irregular areas, and figure your overhead. **448 pages, 8½ x 11, $39.50**

Markup & Profit: A Contractor's Guide

In order to succeed in a construction business, you have to be able to price your jobs to cover all labor, material and overhead expenses, and make a decent profit. The problem is knowing what markup to use. You don't want to lose jobs because you charge too much, and you don't want to work for free because you've charged too little. If you know how to calculate markup, you can apply it to your job costs to find the right sales price for your work. This book gives you tried and tested formulas, with step-by-step instructions and easy-to-follow examples, so you can easily figure the markup that's right for your business. Includes a CD-ROM with forms and checklists for your use. **320 pages, 8½ x 11, $32.50**

Construction Estimating Reference Data

Provides the 300 most useful manhour tables for practically every item of construction. Labor requirements are listed for sitework, concrete work, masonry, steel, carpentry, thermal and moisture protection, door and windows, finishes, mechanical and electrical. Each section details the work being estimated and gives appropriate crew size and equipment needed. Includes a CD-ROM with an electronic version of the book with *National Estimator*, a stand-alone *Windows*™ estimating program, plus an interactive multimedia video that shows how to use the disk to compile construction cost estimates. **432 pages, 11 x 8½, $39.50**

How to Succeed With Your Own Construction Business

Everything you need to start your own construction business: setting up the paperwork, finding the work, advertising, using contracts, dealing with lenders, estimating, scheduling, finding and keeping good employees, keeping the books, and coping with success. If you're considering starting your own construction business, all the knowledge, tips, and blank forms you need are here. **336 pages, 8½ x 11, $28.50**

Residential Steel Framing Guide

Steel is stronger and lighter than wood — straight walls are guaranteed — steel framing will not wrap, shrink, split, swell, bow, or rot. Here you'll find full page schematics and details that show how steel is connected in just about all residential framing work. You won't find lengthy explanations here on how to run your business, or even how to do the work. What you will find are over 150 easy-to-read full-page details on how to construct steel-framed floors, roofs, interior and exterior walls, bridging, blocking, and reinforcing for all residential construction. Also includes recommended fasteners and their applications, and fastening schedules for attaching every type of steel framing member to steel as well as wood. **170 pages, 8½ x 11, $38.80**

Contractor's Guide to the Building Code Revised

This new edition was written in collaboration with the International Conference of Building Officials, writers of the code. It explains in plain English exactly what the latest edition of the *Uniform Building Code* requires. Based on the 1997 code, it explains the changes and what they mean for the builder. Also covers the *Uniform Mechanical Code* and the *Uniform Plumbing Code*. Shows how to design and construct residential and light commercial buildings that'll pass inspection the first time. Suggests how to work with an inspector to minimize construction costs, what common building shortcuts are likely to be cited, and where exceptions may be granted. **320 pages, 8½ x 11, $39.00**

Basic Concrete Engineering for Builders

Basic concrete design principles in terms readily understood by anyone who has poured and finished site-cast structural concrete. Shows how structural engineers design concrete for buildings – foundations, slabs, columns, walls, girders, and more. Tells you what you need to know about admixtures, reinforcing, and methods of strengthening concrete, plus tips on field mixing, transit mix, pumping, and curing. Explains how to design forms for maximum strength and to prevent blow-outs, form and size slabs, beams, columns and girders, calculate the right size and reinforcing for foundations, figure loads and carrying capacities, design concrete walls, and more. Includes a CD-ROM with a limited version of an engineering software program to help you calculate beam, slab and column size and reinforcement. **256 pages, 8½ x 11, $39.50**

Getting Financing & Developing Land

Developing land is a major leap for most builders – yet that's where the big money is made. This book gives you the practical knowledge you need to make that leap. Learn how to prepare a market study, select a building site, obtain financing, guide your plans through approval, then control your building costs so you can ensure yourself a good profit. Includes a CD-ROM with forms, checklists, and a sample business plan you can customize and use to help you sell your idea to lenders and investors. **232 pages, 8½ x 11, $39.00**

 Craftsman Book Company
6058 Corte del Cedro, P.O. Box 6500
Carlsbad, CA 92018

☎ 24 hour order line
1-800-829-8123
Fax (760) 438-0398

Order online
http://www.craftsman-book.com
Free on the Internet!
Download any of Craftsman's estimating costbooks for a 30-day free trial! http://costbook.com

Name _____

Company _____

Address _____

City/State/Zip _____

○ This is a residence

Total enclosed_____(In California add 7% tax)
We pay shipping when your check covers your order in full.

In A Hurry?
We accept phone orders charged to your
○ Visa, ○ MasterCard, ○ Discover or ○ American Express

Card# _____

Exp. date_____ Initials _____

Tax Deductible: Treasury regulations make these references tax deductible when used in your work. Save the canceled check or charge card statement as your receipt.

10-Day Money Back Guarantee

○ 36.50 Basic Engineering for Builders
○ 14.75 Blueprint Reading for Building Trades
○ 30.50 Builder's Guide to Accounting Rev.
○ 68.50 CD Estimator
○ 69.00 CD Estimator — Heavy
○ 39.50 Basic Concrete Engineering for Builders
○ 39.50 Construction Estimating Reference Data with FREE *National Estimator* on a CD-ROM.
○ 41.75 Construction Forms & Contracts with a CD-ROM for *Windows*™ and Mac.
○ 39.00 Contractor's Guide to the Building Code Revised
○ 22.25 Contractor's Survival Manual
○ 39.50 Estimating Excavation
○ 39.00 Getting Financing & Developing Land

○ 59.95 Greenbook Standard Specifications for Public Works Construction
○ 28.50 How to Succeed w/Your Own Construction Business
○ 32.50 Markup & Profit: A Contractor's Guide
○ 28.50 Masonry & Concrete Construction Revised
○ 47.50 National Construction Estimator with FREE *National Estimator* on a CD-ROM.
○ 29.00 Pipe & Excavation Contracting
○ 38.80 Residential Steel Framing Guide
○ 22.75 Excavation & Grading Handbook Revised
○ FREE Full Color Catalog

Prices subject to change without notice